Siegfried Gottwald

Fuzzy Sets and Fuzzy Logic

Artificial Intelligence

Künstliche Intelligenz

edited by Wolfgang Bibel and Walther von Hahn

Artificial Intelligence aims for an understanding and the technical realization of intelligent behaviour. The books of this series are meant to cover topics from the areas of knowledge processing, knowledge representation, expert systems, communication of knowledge (language, images, speach, etc.), AI machinery as well as languages, models of biological systems, and cognitive modelling.

In English:

Automated Theorem Proving
by Wolfgang Bibel

Parallelism in Logic
by Franz Kurfeß

Relative Complexities of First Order Calculi
by Elmar Eder

Fuzzy Sets and Fuzzy Logic
Foundations of Application - from a Mathematical Point of View
by Siegfried Gottwald

In German:

Die Wissensrepräsentationssprache OPS 5
by Reinhard Krickhahn and Bernd Radig

Prolog
by Ralf Cordes, Rudolf Kurse, Horst Langendörfer and Heinrich Rust

LISP
by Rüdiger Esser and Elisabeth Feldmar

Logische Grundlagen der Künstlichen Intelligenz
by Michael R. Genesereth and Nils J. Nilsson

Wissensbasierte Echtzeitplanung
by Jürgen Dorn

Modulare Regelprogrammierung
by Siegfried Bocionek

Automatisierung von Terminierungsbeweisen
by Christoph Walther

Logische und Funktionale Programmierung
by Ulrich Furbach

Schließen bei unsicherem Wissen in der Künstlichen Intelligenz
by Léa Sombé

Wissensbasierte Systeme
by Doris Altenkrüger and Winfried Büttner

Siegfried Gottwald

Fuzzy Sets and Fuzzy Logic

The Foundations of Application - from a Mathematical Point of View

Die Deutsche Bibliothek - CIP-Einheitsaufnahme

Gottwald, Siegfried:
Fuzzy sets and fuzzy logic / Siegfried Gottwald. - Braunschweig
; Wiesbaden : Vieweg, 1993
(Artificial Intelligence)
ISBN-13: 978-3-322-86813-8 e-ISBN-13: 978-3-322-86812-1
DOI: 10.1007/978-3-322-86812-1

AMS Subject Classification: 03E72, 03B52, 04A72, 68N17, 68PO5, 93C42, 94D05

Verlag Vieweg, P.O. Box 5829, D-6200 Wiesbaden

Distribution rights for France by Teknea, Toulouse, Marseille, Barcelona.

ISSN 0940-699

Preface

Quite a long time after its inception in about 1965, the field of fuzzy sets and fuzzy logic was ranked to be some exotic field of research. The main reason was its seeming tendency away from "hard" mathematical modelling and thus – as often supposed by mathematically inclined outsiders – completely away from an acceptable mathematically based theory.

The very recent success with even consumer products involving fuzzy tools causes a rapidly growing interest of the engineering community in this field. But it may also become responsible for a far more extended interest from mathematicians and computer scientists.

The term "fuzzy logic" has undergone an almost continuous change. First it merely referred to tools from many-valued logic appearing in considerations of fuzzy sets and sometimes of a generalized switching theory. Later it referred to the field of fuzzy logic control with its limited use of nonstandard reasoning by interpreting implication-like formulations of the rule oriented IF...THEN - type as fuzzy relations between the values of fuzzy variables. And the presently prevailing understanding is that of a theory of approximate reasoning (sometimes including the pure theory of fuzzy sets too).

Therefore it may not be without some interest to give here a reconsideration of a line of research which tends toward a combination of two different mainstreams. One of them is concerned with mathematical, in particular set theoretical and logical tools which are related to problems of a foundational character. The other one originates from problems which have arisen out of applicational, engineering discussions about modelling strategies in the fuzzy field, especially in fuzzy (logic) control. The unifying background idea is the use of notions and methods from many-valued logics in the discussion of problems which, in the end, are related to fuzzy sets applications.

The author's lasting interest in this field arose out of his early work on the set theoretic foundation for fuzzy sets, cf. GOTTWALD (1971, 1971a, 1974, 1976, 1979, 1980) and the historical reminiscences in GOTTWALD (1984c), but was essentially stimulated since the late 1970s by fruitful contacts with (mainly Polish) control engineering people, among whom Witold Pedrycz was the most influential.

The present book gives a rethought, revised and reorganized, as well as rewritten and extended collection of the author's essential results obtained since that time and within, as well as via, this cooperation – results which directly and indirectly are related to fuzzy relation equations and fuzzy control.

Chapter 1 is essentially based on the research reported in GOTTWALD (1984), but contains also ideas of GOTTWALD (1986, 1986a). Chapter 2 is based on GOTTWALD (1986a), extends those results partly with results from GOTTWALD (1974, 1986) and also includes unpublished material announced in GOTTWALD (1991). Chapter 3 is based on GOTTWALD (1983, 1986, 1992), as well as on parts of GOTTWALD/PEDRYCZ (1986, 1988). Chapter 4 combines results from GOTTWALD (1984a, 1986b) and GOTTWALD/PEDRYCZ (1986) with newly written (and yet unpublished) parts. The final chapter 5 has GOTTWALD/PEDRYCZ (1986, 1986a, 1988) as its most essential sources.

The final writing of this book was initiated and decisively made possible by a grant from the SEL-Stiftung (Stuttgart) that enabled the author to stay for the period of writing away from his home university of Leipzig at the Computer Science Department of the Technische Hochschule Darmstadt within the stimulating atmosphere of the FG Intellektik, whose chairman Wolfgang Bibel had the final responsibility for the author's excellent working conditions.

To all these institutions and those – named as well as unnamed – people the author is exceptionally grateful. He also is grateful to his wife and family for their stimulating interest, as well as for their success in leaving him untouched by almost all the small problems of everyday life.

Darmstadt / Leipzig,
Summer 1992.

Contents

Chapter 1

Logical Preliminaries

1.1 Basic notions

A *fuzzy set A* over some universe of discourse $\mathcal{X}$ (which itself is a classical set) is characterized by its *membership function* $\mu_A : \mathcal{X} \to [0,1]$ and often called *fuzzy subset* of $\mathcal{X}$. The fuzzy sets will be identified with their membership functions such that the class $\mathbb{F}(\mathcal{X})$ of all fuzzy subsets of the universe of discourse $\mathcal{X}$ becomes the class $[0,1]^{\mathcal{X}}$ of all functions f with $\mathrm{dom}(f) = \mathcal{X}$ and $\mathrm{rg}(f) \subseteq [0,1]$.

Fuzzy sets are thus generalized characteristic functions. In essentially the same way as the values $\chi_{\mathcal{M}}(a)$ of the characteristic function $\chi_{\mathcal{M}}$ of a classical, "crisp" set $\mathcal{M}$ code the truth and falsity of the predicate "$a \in \mathcal{M}$" such that

$$\chi_{\mathcal{M}}(a) = \begin{cases} 1, & \text{iff } a \in \mathcal{M}, \quad \text{i.e. iff "} a \in \mathcal{M} \text{" is true} \\ 0, & \text{iff } a \notin \mathcal{M}, \quad \text{i.e. iff "} a \in \mathcal{M} \text{" is false,} \end{cases}$$

the membership degrees of fuzzy sets are considered as generalized truth values, i.e. as *truth degrees* of a suitable many-valued logic **L**.

The many-valued logic **L** which we will use has the real interval [0,1] as its set of truth degrees and is supposed to be *semantically based,* i.e. we assume that the connectives and quantifiers of the language of **L** will be defined via truth functions and truth functional conditions, and we assume that the truth degree 1 is the only (positively) designated one. We will not look at problems of axiomatization here.

The truth functions of many-valued logic which are chosen to define e.g. a conjunction, a disjunction, or a negation are usually supposed to behave in such a way that they assume a value out of $\{0,1\}$ if all their arguments are from $\{0,1\}$ – and additionally to coincide under such circumstances with the binary truth functions of conjunction, disjunction, and negation of classical (propositional) logic. But other, more nonstandard truth functions are possible and have been considered; cf. RESCHER (1969), GOTTWALD (1989).

Many-valued logic as a separate field of logical investigations essentially developed out of the work of the Polish logician J. ŁUKASIEWICZ and was extensively developed in Poland in the 1920s; cf. ŁUKASIEWICZ/TARSKI (1930), ŁUKASIEWICZ (1970). Since then, the real unit interval [0,1] has traditionally been a standard choice for the set of truth degrees in many-valued logic. And the functions min, max are the truth functions of standard generalizations of conjunction and disjunction.

As usual we denote with $\wedge$ the (generalized) *conjunction* with truth function min : $[0,1]^2 \rightarrow [0,1]$ and denote with $\vee$ the (generalized) *disjunction* with truth function max : $[0,1]^2 \rightarrow [0,1]$. Additionally a kind of standard (generalized) *negation* $\neg$ with truth function $n_L(x) = 1 - x$ usually is considered.

Yet, whole classes of further truth functions for conjunctions, disjunctions, implications, and negations seem to be of interest. Therefore, as connectives we furthermore suppose to have conjunctions $\wedge_t$ defined by given t-norms t and implication operators $\rightarrow_t$ also defined by given t-norms t (but via so-called Φ-operators connected with the t-norms t); additionally we connect with each t-conorm s_t a disjunction operator $\vee_t$ and with every negation function n a negation operator $\sim_n$. (Those types of truth functions we have referred to will be defined in the next section.)

As far as quantifiers are concerned, we take one for many-valued universal quantification (denoted: $\forall$) and another one for many-valued existential quantification (denoted: $\exists$).

The well-formed formulas of our language are supposed to be defined in the usual way. And, assuming that the actual values of all free variables of a well-formed formula H always are determined by the context, we write $[\![H]\!]$ for the truth degree of this formula H.

For the quantifiers we have independent of the t-norms under consideration

$$[\![\forall x H(x)]\!] = \inf_{a \in \mathcal{X}} [\![H(x/a)]\!], \tag{1.1}$$

$$[\![\exists x H(x)]\!] = \sup_{a \in \mathcal{X}} [\![H(x/a)]\!] \tag{1.2}$$

where obviously the "substitution notation" $H(x/a)$ does mean that the free variable x of H has to be given the value a from the universe $\mathcal{X}$.

As in classical logic it is often helpful to have besides those unrestricted quantifiers also the restricted ones, usually written $\forall_C x H(x)$ and $\exists_C x H(x)$. Here C indicates some unary predicate or some (crisp) set, and accordingly those restricted quantifications are understood as $\forall x(C(x) \rightarrow H(x))$ or $\exists x(C(x) \wedge H(x))$ for a predicate C and as $\forall x(x \in C \rightarrow H(x))$ or $\exists x(x \in C \wedge H(x))$ for a set C.

In many-valued logic one has to be more careful in that translation. Principally one ought to distinguish two cases: C marking a classical, two-valued restriction (realized as a unary "crisp" predicate or a crisp set) or C marking a "many-valued restriction". It seems that for the latter case a good intuitive understanding is lacking such that in such a case one should prefer to work with unrestricted quantifications. But if C is a crisp restriction now taken as a crisp set, a quite natural understanding of restricted quantifications is at hand:

$$[\![\forall_C x H(x)]\!] = \inf_{a \in C} [\![H(x/a)]\!], \tag{1.3}$$

$$[\![\exists_C x H(x)]\!] = \sup_{a \in C} [\![H(x/a)]\!]. \tag{1.4}$$

(If C means a "crisp" predicate, substitute $C(a)$ for $a \in C$.)

The language of **L** is thus supposed to be a first order one. It is not necessary to have all the details of this language fixed in advance. We will – from time to time, if necessary – change some details without causing confusion by this; but generally, for the set theoretic applications which later on become of central importance, we suppose there are two sorts of variables availabe:

– lower case Latin letters like $a, b, c, \ldots, x, y, z$ as constants and variables for elements of the universe of discourse $\mathcal{X}$;

– upper case Latin letters like $A, B, C, \ldots, X, Y, Z$ as constants and variables for fuzzy subsets of some universe of discourse $\mathcal{X}$.

The basic binary predicate symbol that denotes the membership relation of elements of the universe of discourse $\mathcal{X}$ with respect to some fuzzy subset of $\mathcal{X}$ will be "ε" and the atomic formulas built up with it are of type "$a \,\varepsilon\, A$". Their truth degrees correspond to the membership degrees of the fuzzy sets, such that for each fuzzy set $A \in \mathbb{F}(\mathcal{X})$ and each $x \in \mathcal{X}$ one has the fundamental equation

$$[\![x \,\varepsilon\, A]\!] =_{def} \mu_A(x). \tag{1.5}$$

If necessary, we will add variables and constants for fuzzy relations too. Furthermore, we need predicate symbols for some many-valued generalizations of usual set-theoretic relations. These predicate symbols will have almost the same graphical shape for fuzzy sets as have the analogous ones for crisp sets; by use of ε they will normally be introduced through definitions in the same style as is usually done in classical set theory.

In a few cases we are interested to have (propositional) constants for the truth degree 1, i.e. for any logically valid expression, as well as for the truth degree 0. We use $\top$ as a constant for the truth degree 1 and $\bot$ as a constant for the truth degree 0.

Our choice of truth degree 1 as the only designated one has as a consequence that we consider a formula H as (*logically*) *valid*, denoted: $\models H$, iff H has (for all values of its free variables) truth degree 1:

$$\models H \quad \text{iff} \quad [\![H]\!] = 1. \tag{1.6}$$

And a further consequence of this choice is that a rule of inference

$$\frac{H_1, \dots, H_n}{H}$$

with the premises $H_1, \dots, H_n$ and the conclusion H is *correct* iff H has truth degree 1 in each situation in which all the premisses $H_1, \dots, H_n$ have truth degree 1.

Obvious examples of correct rules of inference are the *rule of generalization*

$$\frac{H(x)}{\forall x H(x)} \tag{1.7}$$

and the *rule of particularization*

$$\frac{H(x/a)}{\exists x H(x)} \tag{1.8}$$

for first order many-valued logic. The correctness proofs here consist in the simple remarks that

$$\text{if } \models H(x) \quad \text{then } \models \forall x H(x)$$

as well as

$$\text{if } \models H(x/a) \quad \text{then } \models \exists x H(x)$$

hold true for every well-formed formula H with x among its free variables and any element a of the corresponding universe.

Of course, the rule of generalization also holds true for each possible restricted universal quantification (1.3) and the rule of particularization for all those restricted existential quantifications (1.4) of the form "$\exists_C x$" with $a \in C$.

To avoid too many parentheses, the usual kind of ranking of the binding strength of the logical constants will be used: the (restricted as well as unrestricted) quantifiers bind stronger than the negations, which bind stronger than the conjunctions and disjunctions, and those connectives finally bind stronger than any implication or biimplication connective.

In our classical metalanguage we use $\Rightarrow$ for (material) implication, $\Leftrightarrow$ for biimplication, and sometimes $\bigwedge$ for universal quantification and $\bigvee$ for existential quantification.

1.2 t-norms and Φ-operators

Definition 1.1 (a) *A binary operation* $\mathbf{t}$ *in the real interval* $[0,1]$ *is a* t-norm *iff it is*

(T1) *associative and commutative;*

(T2) *non-decreasing in the first – and hence in each – argument;*

(T3) *has* 1 *as neutral element, i.e.* $u \,\mathbf{t}\, 1 = u$ *for each* $u \in [0,1]$.

(b) *A binary operation* $\mathbf{s}$ *in* $[0,1]$ *is a* t-conorm *iff it is*

(S1) *associative and commutative;*

(S2) *non-decreasing in each argument;*

(S3) *has* 0 *as neutral element, i.e.* $u \,\mathbf{s}\, 0 = u$ *for each* $u \in [0,1]$.

The t-norms – a common shorthand term for "triangular norms" – have been widely used in investigations into probabilistic metric spaces, cf. e.g. SCHWEIZER/SKLAR (1961, 1983), LING (1965), FRANK (1979). From those investigations and also independent of them, it can be argued that the t-norms are suitable candidates for conjunctions in many-valued logic, and the t-conorms candidates for disjunctions. This point of view is explained in the next section.

Concerning notation we always will feel free, for t-norms as well as t-conorms and later on for Φ-operators too, to use them either in prefix or in infix notation, i.e. to take them either as two-place functions in [0,1] or as binary operations in [0,1]. That means for a t-norm $\boldsymbol{t}$ that one always has: $u\,\boldsymbol{t}\,v = \boldsymbol{t}(u,v)$, and that the type of notation is changed without further comment according to which notation seems to be best in the particular context.

In Table 1.1 we will now list some t-norms which are usually main examples or which are members of well known families of t-norms.

t-norm $\boldsymbol{t}$	formula for $\boldsymbol{t}(u,v)$
$\boldsymbol{t}_G$	$\min\{u,v\}$
$\boldsymbol{t}_L$	$\max\{0, u+v-1\}$
$\boldsymbol{t}_P$	$u \cdot v$
$\boldsymbol{t}_D$	$\begin{cases} \min\{u,v\} & \text{for } u=1 \\ & \text{or } v=1 \\ 0 & \text{otherwise} \end{cases}$
$\boldsymbol{t}^{\mathrm{H}}_{\gamma}, \quad \gamma \geq 0$	$\dfrac{uv}{\gamma + (1-\gamma)\cdot(u+v-uv)}$
$\boldsymbol{t}^{\mathrm{Y}}_{p}, \quad p > 0$	$1 - \min\{((1-u)^p + (1-v)^p)^{1/p}, 1\}$
$\boldsymbol{t}^{\mathrm{D}}_{\lambda}, \quad \lambda > 0$	$1 - \dfrac{1}{1 + ((\frac{1-u}{u})^\lambda + (\frac{1-v}{v})^\lambda)^{1/\lambda}}$
$\boldsymbol{t}^{\mathrm{W}}_{\lambda}, \quad \lambda > -1$	$\max\{0, \dfrac{u+v-1+\lambda uv}{1+\lambda}\}$

Table 1.1: Examples of t-norms[1]

Generally, the t-norms and the t-conorms can be reduced one to another.

[1] The last four families of t-norms each aim to cover a whole "interval" of possible t-norms between borderline cases given by the extremal values of the respective parameters. They have been introduced by HAMACHER (1978), YAGER (1980), DOMBI (1982) and WEBER (1983). The upper index in each case points to the name of the defining author. Pictorial representations of those t-norms and even more families of t-norms are presented in MIZUMOTO (1989).

For, given any t-norm $\boldsymbol{t}$, by

$$\boldsymbol{s}_{\boldsymbol{t}}(u,v) =_{def} 1 - \boldsymbol{t}(1-u, 1-v) \tag{1.9}$$

a t-conorm $\boldsymbol{s}_{\boldsymbol{t}}$ is defined; and given any t-conorm $\boldsymbol{s}$ by

$$\boldsymbol{t}_{\boldsymbol{s}}(u,v) =_{def} 1 - \boldsymbol{s}(1-u, 1-v) \tag{1.10}$$

a t-norm $\boldsymbol{t}_{\boldsymbol{s}}$ is defined. Furthermore, it is easy to see that this is a 1-1 correspondence between t-norms and t-conorms because of

$$(\boldsymbol{t})_{\boldsymbol{s}_{\boldsymbol{t}}} = \boldsymbol{t}, \qquad (\boldsymbol{s})_{\boldsymbol{t}_{\boldsymbol{s}}} = \boldsymbol{s} \tag{1.11}$$

for each t-norm $\boldsymbol{t}$, t-conorm $\boldsymbol{s}$.

As a consequence of (1.11) one can either focus attention on the t-norms or on the t-conorms: in any case it is not necessary to treat both types of functions *à par*. We have decided here to take the t-norms to be of greater importance. Hence, in the following all the t-conorms $\boldsymbol{s}$ will be considered as t-conorms of the form $\boldsymbol{s}_{\boldsymbol{t}}$ defined via some t-norm $\boldsymbol{t}$.

For the t-norms, as specific real functions, one has the natural pointwise partial ordering of functions which as a reflexive ordering is defined by

$$\boldsymbol{t}_1 \leqq \boldsymbol{t}_2 \quad =_{def} \quad \boldsymbol{t}_1(u,v) \leq \boldsymbol{t}_2(u,v) \;\text{ for all }\; u,v \in [0,1] \tag{1.12}$$

for any t-norms $\boldsymbol{t}_1, \boldsymbol{t}_2$ and as an irreflexive ordering by

$$\boldsymbol{t}_1 < \boldsymbol{t}_2 \quad =_{def} \quad \boldsymbol{t}_1 \leqq \boldsymbol{t}_2 \;\text{ and }\; \boldsymbol{t}_1 \neq \boldsymbol{t}_2. \tag{1.13}$$

Because from the defining properties (T2), (T3) one has $0 \leq u\,\boldsymbol{t}\,v \leq u\,\boldsymbol{t}\,1 = u$ as well as $0 \leq u\,\boldsymbol{t}\,v \leq 1\,\boldsymbol{t}\,v = v$ for any t-norm $\boldsymbol{t}$ and all $u, v \in [0,1]$, that means all together that

$$\boldsymbol{t}_D \leqq \boldsymbol{t} \leqq \boldsymbol{t}_G \qquad \text{for each t-norm } \boldsymbol{t}. \tag{1.14}$$

Thus for the first four examples of t-norms from Table 1.1 there holds furthermore

$$\boldsymbol{t}_D < \boldsymbol{t}_L < \boldsymbol{t}_P < \boldsymbol{t}_G \tag{1.15}$$

as can be seen almost directly.

The same partial orderings (1.12), (1.13) also apply to t-conorms. And for t-conorms their orderings are dual to those for the corresponding t-norms in the sense that for all t-norms $\boldsymbol{t}_1, \boldsymbol{t}_2$ one has

$$\boldsymbol{s}_{\boldsymbol{t}_1} \leqq \boldsymbol{s}_{\boldsymbol{t}_2} \;\Leftrightarrow\; \boldsymbol{t}_2 \leqq \boldsymbol{t}_1 \tag{1.16}$$

and hence also

$$s_{t_1} < s_{t_2} \Leftrightarrow t_2 < t_1.$$

Therefore it suffices to consider these orderings for t-norms only.

Of course, equations (1.9) and (1.10) correspond to the usual DE MORGAN laws which connect conjunction and disjunction: the essential point is that here the function $x \mapsto 1 - x$ has to be interpreted as describing a generalized negation, but really it is the truth function of the "standard" negation $\neg$.

This function $x \mapsto 1 - x$ is not the only candidate for a generalized negation; but it is the most commonly considered one. Yet, e.g. WEBER (1983) and KLEMENT (1982) discuss other negation functions too. Following WEBER (1983) we give

Definition 1.2 *A unary operation n in the real interval* $[0, 1]$ *is a* negation function *iff the following hold true*

(N1) $n(0) = 1, \quad n(1) = 0,$

(N2) *n is non-increasing;*

additionally such a negation function will be called strict *iff*

(N3) *n is decreasing and continuous,*

finally n is an involution *iff n is a strict negation function and fulfills*

(N4) $n(n(u)) = u$ *for each* $u \in [0, 1]$.

Clearly, the function $x \mapsto 1 - x$ is an involution. Furthermore, the connections of t-norms and t-conorms established in equations (1.9), (1.10) can be generalized to any strict negation function. WEBER (1983) proved that for any t-norm t and strict negation function n by

$$s_{n,t}(u, v) = n^{-1}(t(n(u), n(v))) \tag{1.17}$$

a t-conorm is defined; and that for any t-conorm s and strict negation function n by

$$t_{n,s}(u, v) = n^{-1}(s(n(u), n(v))) \tag{1.18}$$

a t-norm is defined. For involutions n in this way a 1-1 correspondence is established between t-norms and t-conorms.

For the many-valued connectives having these t-norms, t-conorms or negation functions as truth functions, clearly equations (1.17) and (1.18) validate corresponding DEMORGAN laws.

t	$s_t(u,v)$
t_G	$\max\{u,v\}$
t_L	$\min\{1, u+v\}$
t_P	$u+v-u\cdot v$
t_D	$\begin{cases} \max\{u,v\} & \text{for } u\cdot v=0 \\ 1 & \text{otherwise} \end{cases}$
$t^{\mathrm{H}}_{\gamma}, \quad \gamma \geq 0$	$\dfrac{u+v-uv-(1-\gamma)uv}{1-(1-\gamma)uv}$
$t^{\mathrm{Y}}_{p}, \quad p > 0$	$\min\{(u^p+v^p)^{1/p}, 1\}$
$t^{\mathrm{D}}_{\lambda}, \quad \lambda > 0$	$\dfrac{1}{1+((\frac{u}{1-u})^{\lambda}+(\frac{v}{1-v})^{\lambda})^{-1/\lambda}}$
$t^{\mathrm{W}}_{\lambda}, \quad \lambda > -1$	$\min\{u+v+\lambda uv, 1\}$

Table 1.2: Examples of t-norms and their t-conorms

In Table 1.2 we have listed for the t-norms of Table 1.1 the t-conorms related to them via (1.9).

Besides the truth functions for conjunction, disjunction and negation in many-valued logic we will have to consider truth functions for implications. That role will be played by the Φ-operators. In section 1.3 we will discuss logically valid formulas of a language of many-valued logic which has just these types of truth functions interpreting their propositional connectives. The results there will indicate that the t-norms are suitable candidates for generalized, i.e. many-valued conjunction operators and that the Φ-operators are equally suitable candidates for many-valued implication operators.

Definition 1.3 *A binary operation* φ *in the real unit interval* $[0,1]$ *is called* Φ-operator *(connected with a given t-norm* **t***) iff for all* $u,v,w \in [0,1]$ *the following hold true*

(Φ1) $v \leq w \Rightarrow u\,\varphi\,v \leq u\,\varphi\,w$;

(Φ2) $u\,\mathbf{t}\,(u\varphi v) \leq v$;
(Φ3) $v \leq u\varphi(u\,\mathbf{t}\,v)$.

The Φ-operators have been introduced by PEDRYCZ (1982) to describe solutions of fuzzy equations which are formulated using the t-norm which the Φ-operator is connected with. They generalize the α-operation of SANCHEZ (1976, 1977), which now becomes the special case of the Φ-operator connected with the t-norm min.

As a side remark let us mention that the definition of Φ-operators as given by PEDRYCZ (1982) used the condition

$$\max\{u\,\varphi\,v, u\,\varphi\,w\} \leq u\,\varphi \max\{v, w\}$$

which obviously is equivalent to our (Φ1).

The character of the connection between a Φ-operator and its corresponding t-norm is not made completely clear by (Φ1) to (Φ3). It was in GOTTWALD (1984, 1986) where this connection was proved in full generality. The results are given in the following propositions 1.1 and 1.3.

Proposition 1.1 (Representation Lemma) *If φ is a Φ-operator which is connected to the t-norm $\mathbf{t}$, then for all $u, v \in [0,1]$ it holds true that*

$$u\,\varphi\,v = \sup\{w \mid u\,\mathbf{t}\,w \leq v\}. \tag{1.19}$$

Proof. By (Φ2) we immediately get

$$u\,\varphi\,v \leq \sup\{w \mid v\,\mathbf{t}\,w \leq v\}.$$

But in case $u\,\varphi\,v < \sup\{w \mid v\,\mathbf{t}\,w \leq v\}$ there would exist w_0 such that $u\,\varphi\,w_0$ and $u\,\mathbf{t}\,w_0 \leq v$, therefore

$$w_0 \leq u\,\varphi(u\,\mathbf{t}\,w_0) \leq u\,\varphi\,v < w_0,$$

which is a contradiction. Thus our representation lemma is proved. QED

Therefore, one Φ-operator at most is connected to each t-norm, which in case of existence can be defined using formula (1.19).

Furthermore, the supremum in (1.19) really is a maximum because by (Φ2) we have $u\,\mathbf{t}\,(u\,\varphi\,v) \leq v$ and thus

$$u\,\mathbf{t}\,\sup\{w \mid u\,\mathbf{t}\,w \leq v\} = u\,\mathbf{t}\,(u\,\varphi\,v) \leq v.$$

Hence it is always even

$$u \varphi v = \max\{w \mid u \, \boldsymbol{t} \, w \leq v\}.$$

But what about the existence of a Φ-operator for a given t-norm? Does such a Φ-operator always exist, i.e. is the binary operation φ defined via (1.19) a Φ-operator for each t-norm? To formulate the answer we need an additional property of t-norms, their left continuity or their lower semicontinuity (in each of the arguments).

Because of the commutativity (T1) of the t-norms it is enough to consider these properties for the first argument only. The *left continuity* of a t-norm $\boldsymbol{t}$ is then, of course, the condition that for all $u_0, v_0 \in [0,1]$ and all convergent sequences $(u_i)_{i \geq 1}$ of points from $[0,1]$ with $\lim_{i \to \infty} u_i = u_0$ and always $u_i < u_0$ one has

$$\lim_{i \to \infty} (u_0 \, \boldsymbol{t} \, v_0) = u_0 \, \boldsymbol{t} \, (\lim_{i \to \infty} u_i).$$

And the *lower semicontinuity* of $\boldsymbol{t}$ means that for each $u_0, v_0 \in [0,1]$ and each $\epsilon > 0$ there is a $\delta > 0$ such that $u \, \boldsymbol{t} \, v_0 > u_0 \, \boldsymbol{t} \, v_0 - \epsilon$ for all $u \in (u_0 - \delta, u_0]$.

Both these notions coincide for t-norms.

Proposition 1.2 *A t-norm is left continuous iff it is lower semicontinuous.*

Proof. More general, for each monotonically non-decreasing function the lower semicontinuity and the left continuity coincide. Hence suppose that f is monotonically non-decreasing.

Assume first that f is lower semicontinuous, $x \in \operatorname{dom}(f)$ and $(x_i)_{i<\infty}$ a converging sequence of points $x_i \in \operatorname{dom}(f)$ with $x_i < x$ for all i. Then $\lim_{i \to \infty} f(x_i)$ exists and one has $\lim_{i \to \infty} f(x_i) \leq f(x)$. In the case that $\lim_{i \to \infty} f(x_i) < f(x)$ let $2\epsilon = f(x) - \lim_{i \to \infty} f(x_i)$. Then in each neighbourhood of x there is some x_k with $f(x_k) < f(x) - \epsilon$ contradicting the lower semicontinuity of f. Thus $\lim_{i \to \infty} f(x_i) = f(x)$ and thus f is left continuous (at x and hence in the whole domain).

Now suppose that f is left continuous. Consider $x_0 \in \operatorname{dom}(f)$ and some $\epsilon > 0$. Then there exists some $\delta > 0$ such that $|f(x_0) - f(x)| < \epsilon$ for all $x \in \operatorname{dom}(f)$ with $x_0 - x < \delta$. By monotonicity of f hence $f(x_0) - f(x) < \epsilon$, i.e. $f(x) > f(x_0) - \epsilon$ and thus f is lower semicontinuous (at x_0 and hence in $\operatorname{dom}(f)$).

Because each t-norm by (T2) is monotonically non-decreasing in each argument the lower semicontinuity of a t-norm (in one and hence both of its arguments) coincides with their left continuity. QED

In the following we write

$$\mathsf{LSC}(\boldsymbol{t}) \quad \text{for} \quad \boldsymbol{t} \text{ is lower semicontiuous.}$$

The characterization of the lower semicontinuity for t-norms we most often need later on is a simple corollary of the last proof and reads:

$$\mathsf{LSC}(\boldsymbol{t}) \quad \Leftrightarrow \quad u\,\boldsymbol{t}\,(\sup_{\xi\in\Xi} v_\xi) = \sup_{\xi\in\Xi}(u\,\boldsymbol{t}\,v_\xi) \quad \text{for all } u, (v_\xi)_{\xi\in\Xi} \text{ from } [0,1]. \tag{1.20}$$

In a few cases it is not the lower semicontinuity of a t-norm $\boldsymbol{t}$ that is needed, but the upper semicontinuity[2] or, which again is the same, the right continuity of $\boldsymbol{t}$. (This equivalence can be proven as in the last proof.) We write $\mathsf{USC}(\boldsymbol{t})$ if $\boldsymbol{t}$ is upper semicontinuous. Again, the crucial property we usually need in case $\mathsf{USC}(\boldsymbol{t})$ is that

$$\mathsf{USC}(\boldsymbol{t}) \quad \Leftrightarrow \quad u\,\boldsymbol{t}\,(\inf_{\xi\in\Xi} v_\xi) = \inf_{\xi\in\Xi}(u\,\boldsymbol{t}\,v_\xi) \quad \text{for all } u, (v_\xi)_{\xi\in\Xi} \text{ from } [0,1]. \tag{1.21}$$

Now we are able to solve the problem of the existence of a Φ-operator with respect to a t-norm $\boldsymbol{t}$.

Proposition 1.3 (Existence Lemma) *For a t-norm* $\boldsymbol{t}$ *there exists a* Φ*-operator* φ *connected with* $\boldsymbol{t}$ *iff* $\boldsymbol{t}$ *is lower semicontinuous.*

Proof. Assume $\mathsf{LSC}(\boldsymbol{t})$. Let φ be defined by (1.19). Obviously this binary operator φ in [0,1] is non-decreasing in the second argument, i.e. fulfills (Φ1). By $\mathsf{LSC}(\boldsymbol{t})$ one has always

$$\begin{aligned} u\,\boldsymbol{t}(u\,\varphi\,v) &= u\,\boldsymbol{t}(\sup\{w \mid u\,\boldsymbol{t}\,w \le v\}) \\ &= \sup\{u\,\boldsymbol{t}\,w \mid u\,zt\,w \le v\} \le v \end{aligned}$$

and thus (Φ2). And for (Φ3) one immediately has

$$u\,\varphi(u\,\boldsymbol{t}\,v) = \sup\{w \mid u\,\boldsymbol{t}\,w \le u\,\boldsymbol{t}\,v\} \ge v.$$

Now suppose that $\mathsf{LSC}(\boldsymbol{t})$ fails. Then there exist in $[0,1]$ real numbers u_0, v_0 and a family $(v_\xi)_{\xi\in\Xi}$ of such numbers with

$$v_0 = \sup_{\xi\in\Xi}(u_0\,\boldsymbol{t}\,v_\xi) < u_0\,\boldsymbol{t}\,(\sup_{\xi\in\Xi}(v_\xi)).$$

[2]The *upper semicontinuity* of t means that for each $u_0, v_0 \in [0,1]$ and each $\epsilon > 0$ there is a $\delta > 0$ such that $u\,\mathrm{t}\,v_0 < u_0\,\mathrm{t}\,v_0 + \epsilon$ for all $u \in [u_0, u_0 + \delta)$.

If we assume that there exists a Φ-operator φ connected to $\boldsymbol{t}$ then we have from the representation lemma that

$$u_0\,\boldsymbol{t}\,(u_0\,\varphi\,v_0) = u_0\,\boldsymbol{t}\,\sup\{w \mid u_0\,\boldsymbol{t}\,w \le v_0\} \ge u_0\,\boldsymbol{t}\,(\sup_{\xi\in\Xi}(v_\xi)) > v_0,$$

contradicting (Φ2). Hence in that case where $\mathsf{LSC}(\boldsymbol{t})$ fails there does not exist a Φ-operator connected with $\boldsymbol{t}$. QED

Definition 1.4 *We denote the uniquely determined Φ-operator which is connected via (1.19) with each lower semicontinuous t-norm $\boldsymbol{t}$ by $\varphi_{\boldsymbol{t}}$, i.e. we put*

$$u\,\varphi_{\boldsymbol{t}}\,v = \sup\{w \mid u\,\boldsymbol{t}\,w \le v\}. \tag{1.22}$$

In Table 1.3 we give a list of Φ-operators which correspond via (1.22) to t-norms of Table 1.1. Here the Φ-operator $\varphi_{\boldsymbol{t}_L}$ is the truth function of the well known ŁUKASIEWICZ implication of many-valued logic and $\varphi_{\boldsymbol{t}_G}$ is related to another such implication operator, often named after K. GÖDEL[3].

From formulas (1.19), (1.22) it is easy to prove some fundamental properties of every Φ-operator.

Proposition 1.4 *For each binary operator φ defined by formula (1.19) it holds true for all u, v, w:*

(i) $u \le v \Rightarrow u\,\varphi\,w \ge v\,\varphi\,w$;

(ii) $u \le v \Rightarrow u\,\varphi\,v = 1$;

(iii) $1\,\varphi\,v = v$;

(iv) $v \le u\,\varphi\,v$.

Proof. Obvious from (1.19).

Proposition 1.5 *For every t-norm $\boldsymbol{t}$ with property $\mathsf{LSC}(\boldsymbol{t})$ one has for all u, v, w:*

(i) $u\,\varphi_{\boldsymbol{t}}(v\,\varphi_{\boldsymbol{t}}\,(u\,\boldsymbol{t}\,v)) = 1$;

(ii) $(u\,\boldsymbol{t}\,v)\,\varphi_{\boldsymbol{t}}\,w \le u\,\varphi_{\boldsymbol{t}}\,(v\,\varphi_{\boldsymbol{t}}\,w)$;

(iii) $u\,\varphi_{\boldsymbol{t}}\,v \le (u\,\boldsymbol{t}\,w)\,\varphi_{\boldsymbol{t}}\,(v\,\boldsymbol{t}\,w)$.

[3]The implication operator of many-valued logic with that truth function was introduced in GÖDEL (1932) in connection with his studies on intuitionistic logic. There he also used a negation which was related to his implication by just the same relation we shall use later on in definition 1.8 in general to introduce a negation function for each t-norm.

t	$\varphi_t(u,v)$
t_G	$\begin{cases} 1 & \text{for } u \leq v \\ v & \text{for } u > v \end{cases}$
t_L	$\min\{1, 1-u+v\}$
t_P	$\begin{cases} \min\{1, v/u\} & \text{for } u \neq 0 \\ 1 & \text{for } u = 0 \end{cases}$
$t^H_\gamma, \quad \gamma \geq 0$	$\dfrac{v+(\gamma-1)(1-u)v}{u+(\gamma-1)(1-u)v} \quad$ for $u > v$ [4]
$t^Y_p, \quad p > 0$	$1-((1-v)^p-(1-u)^p)^{1/p} \quad$ for $u > v$
$t^D_\lambda, \quad \lambda > 0$	$\dfrac{1}{1+((\frac{1-v}{v})^\lambda-(\frac{1-u}{u})^\lambda)^{1/\lambda}} \quad$ for $u > v$
$t^W_\lambda, \quad \lambda > -1$	$\dfrac{1-u+v+\lambda v}{1+\lambda u} \quad$ for $u > v$

Table 1.3: Examples of t-norms and their Φ-operators [5]

Proof. (i) is obvious from (Φ2) and proposition 1.4 (ii). To get (ii) we observe that by (1.22) it is enough to prove

$$u\,t\,((u\,t\,v)\,\varphi_t\,w) \leq v\,\varphi_t\,w,$$

and for this again by (1.22) it is enough to prove

$$v\,t\,(u\,t\,((u\,t\,v)\,\varphi_t\,w)) \leq w$$

which is true because of (Φ2) and the associativity and commutativity of t. For (iii) it is, again by (1.22), enough to prove

$$u\,t\,s \leq v \Rightarrow (u\,t\,t)\,t\,s \leq v\,t\,t$$

which is obvious by associativity and monotonicity of t, i.e. by conditions (T1), (T2). QED

[4] Here and in the following lines in case $u \leq v$ the result is always 1 as in the first line.

[5] Of course here the drastic product t_D has to fail because t_D is not lower semicontinuous, i.e. not left continuous for each point $(1, v_0)$ of $[0,1]^2$ with $v_0 < 1$.

Proposition 1.6 *Suppose* $\mathsf{LSC}(\boldsymbol{t})$. *Then it holds true for all* $u, v, w \in [0,1]$:

$$
\begin{array}{ll}
(i) & u \leq v \Leftrightarrow u\,\varphi_{\boldsymbol{t}}\,v = 1; \\
(ii) & (u\,\boldsymbol{t}\,v)\,\varphi_{\boldsymbol{t}}\,w = u\,\varphi_{\boldsymbol{t}}\,(v\,\varphi_{\boldsymbol{t}}\,w); \\
(iii) & (u\,\varphi_{\boldsymbol{t}}\,v)\,\boldsymbol{t}\,(v\,\varphi_{\boldsymbol{t}}\,w) \leq u\,\varphi_{\boldsymbol{t}}\,w.
\end{array}
$$

Proof. (i) Because of proposition 1.4 (i) only ($\Leftarrow$) has to be proved. Hence suppose

$$1 = u\,\varphi_{\boldsymbol{t}}\,v = \sup\{w \mid u\,\boldsymbol{t}\,w \leq v\}.$$

Then by (T2) we get

$$w < 1 \Rightarrow u\,\boldsymbol{t}\,w \leq v$$

and hence

$$u = u\,\boldsymbol{t}\,1 = u\,\boldsymbol{t}\,\sup\{w \mid w < 1\} = \sup\{u\,\boldsymbol{t}\,w \mid w < 1\} \leq v.$$

(ii) Because of proposition 1.5 (ii) it is enough to prove

$$u\,\varphi_{\boldsymbol{t}}\,(v\,\varphi_{\boldsymbol{t}}\,w) \leq (u\,\boldsymbol{t}\,v)\,\varphi_{\boldsymbol{t}}\,w.$$

And by (1.22) for this inequalitiy only

$$u\,\boldsymbol{t}\,s \leq v\,\varphi_{\boldsymbol{t}}\,w \Rightarrow (u\,\boldsymbol{t}\,v)\,\boldsymbol{t}\,s \leq w$$

has to be proved. Hence assume that $u\,\boldsymbol{t}\,s \leq v\,\varphi_{\boldsymbol{t}}\,w$; in this case we have

$$(u\,\boldsymbol{t}\,v)\,\boldsymbol{t}\,s = (v\,\boldsymbol{t}\,u)\,\boldsymbol{t}\,s \leq v\,\boldsymbol{t}\,(v\,\varphi_{\boldsymbol{t}}\,w) \leq w$$

by (T1), (T2) and (Φ2).

(iii) By $\mathsf{LSC}(\boldsymbol{t})$ and formula (1.22) we get in a straightforward manner

$$
\begin{array}{rcl}
(u\,\varphi_{\boldsymbol{t}}\,v)\,\boldsymbol{t}\,(v\,\varphi_{\boldsymbol{t}}\,w) & = & \sup\{s\,\boldsymbol{t}\,t \mid u\,\boldsymbol{t}\,s \leq v \text{ and } v\,\boldsymbol{t}\,t \leq w\} \\
& \leq & \sup\{r \mid u\,\boldsymbol{t}\,r \leq w\} = u\,\varphi_{\boldsymbol{t}}\,w
\end{array}
$$

because of the fact that one has

$$u\,\boldsymbol{t}\,(s\,\boldsymbol{t}\,t) = (u\,\boldsymbol{t}\,s)\,\boldsymbol{t}\,t \leq v\,\boldsymbol{t}\,t \leq w$$

from $u\,\boldsymbol{t}\,s \leq v$ and $v\,\boldsymbol{t}\,t \leq w$. QED

If we even suppose that our t-norms are continuous functions, then the following two interesting additional results become provable.

Proposition 1.7 *For each continuous t-norm* $\boldsymbol{t}$ *the following hold true*

$$(i) \quad u\,\boldsymbol{t}\,(u\,\varphi_{\boldsymbol{t}}\,v) = u \wedge v,$$
$$(ii) \quad v\,\varphi_{\boldsymbol{t}}\,(u\,\boldsymbol{t}\,(u\,\varphi_{\boldsymbol{t}}\,v)) = v\,\varphi_{\boldsymbol{t}}\,u.$$

Proof. (i) We have $u\,\boldsymbol{t}\,(u\,\varphi_{\boldsymbol{t}}\,v) \leq v$ for every u, v and $u\,\boldsymbol{t}\,u \leq u$ for every u. From this and

$$u\,\boldsymbol{t}\,(u\,\varphi_{\boldsymbol{t}}\,v) = \bigvee\{u\,\boldsymbol{t}\,u \mid u\,\boldsymbol{t}\,u \leq v\}$$

for continuous $\boldsymbol{t}$, (i) follows.

(ii) Immediately from (i) we have

$$v\,\varphi_{\boldsymbol{t}}\,(u\,\boldsymbol{t}\,(u\,\varphi_{\boldsymbol{t}}\,v)) = v\,\varphi_{\boldsymbol{t}}\,(u \wedge v) = v\,\varphi_{\boldsymbol{t}}\,u$$

which is obvious for $u \leq v$, but if $v < u$ we have $1 = v\,\varphi_{\boldsymbol{t}}\,v = v\,\varphi_{\boldsymbol{t}}\,u$. QED

Using the Φ-operators we are able to introduce a special kind of negation function with respect to each t-norm $\boldsymbol{t}$ with the property $\mathsf{LSC}(\boldsymbol{t})$ by the following definition.

Definition 1.5 *Suppose* $\mathsf{LSC}(\boldsymbol{t})$. *Then for each* $u \in [0,1]$ *let be*

$$\boldsymbol{n}_{\boldsymbol{t}}(u) =_{def} \varphi_{\boldsymbol{t}}(u, 0).$$

Corollary 1.8 *For each t-norm* $\boldsymbol{t}$ *with the property* $\mathsf{LSC}(\boldsymbol{t})$ *the function* $\boldsymbol{n}_{\boldsymbol{t}} : [0,1] \to [0,1]$ *is a negation function.*

Proof. Obviously $\boldsymbol{n}_{\boldsymbol{t}}(0) = 1$ and $\boldsymbol{n}_{\boldsymbol{t}}(1) = 0$ hold true. And from proposition 1.1 it follows that each such function $\boldsymbol{n}_{\boldsymbol{t}}$ is non-increasing. QED

In general, however, $\boldsymbol{n}_{\boldsymbol{t}}$ need not be a strict negation function or even an involution.

There are some additional properties in the class of t-norms which deserve to be mentioned for the question of the choice of a specific t-norm.

Definition 1.6 *A t-norm* $\boldsymbol{t}$ *is called* idempotent *iff* $u\,\boldsymbol{t}\,u = u$ *for all* $u \in [0,1]$; *and it is called* interactive *iff there exist some* $u_0, v_0 \in [0,1]$ *such that* $u_0\,\boldsymbol{t}\,v_0 \notin \{u_0, v_0\}$. *This terminology is used for t-conorms in the same sense.*

Corollary 1.9 (i) *A t-norm* $\boldsymbol{t}$ *is idempotent iff its corresponding t-conorm* $\boldsymbol{s}_{\boldsymbol{t}}$ *is idempotent.*

(ii) *A t-norm* $\boldsymbol{t}$ *is interactive iff its corresponding t-conorm* $\boldsymbol{s}_{\boldsymbol{t}}$ *is interactive.*

Proof. Both claims are immediate consequences of the corresponding definitions. (For (ii) the simpler proof results if one proves the equivalent claim: $\boldsymbol{t}$ is non-interactive iff $\boldsymbol{s_t}$ is non-interactive.) QED

It is, therefore, enough to discuss t-norms with respect to interactivity and idempotency. The same results then hold true for t-conorms too.

Proposition 1.10 (i) *The only idempotent t-norm is the t-norm* $\boldsymbol{t} = \min$.

(ii) *The only non-interactive t-norm is the t-norm* $\boldsymbol{t} = \min$.

Proof. (i) Suppose that $\boldsymbol{t}$ is an idempotent t-norm. Consider any $u, v \in [0,1]$ and suppose $u \leq v$. Then one has

$$u = u\,\boldsymbol{t}\,u \leq u\,\boldsymbol{t}\,v \leq u\,\boldsymbol{t}\,1 = u$$

by (T2), (T3) and the idempotency of $\boldsymbol{t}$. Therefore $u\,\boldsymbol{t}\,v = \min\{u,v\}$ in this case. And for $v \leq u$ the same result holds true by symmetry.

(ii) Assume that $\boldsymbol{t}$ is non-interactive. Then for each $u \in [0,1]$ one has $u\,\boldsymbol{t}\,u \in \{u\}$ and thus $u\,\boldsymbol{t}\,u = u$. Hence $\boldsymbol{t}$ is idempotent and therefore $\boldsymbol{t} = \min$ by (i). QED

Proposition 1.11 (i) *If a t-norm* $\boldsymbol{t}$ *is distributive with respect to its corresponding t-conorm* $\boldsymbol{s_t}$ *then* $\boldsymbol{t} = \min$.

(ii) *If a t-conorm* $\boldsymbol{s_t}$ *is distributive with respect to its corresponding t-norm* $\boldsymbol{t}$ *then* $\boldsymbol{t} = \min$.

Proof. (i) Assume that this kind of distributivity holds, i.e. that for all $u, v, w \in [0,1]$

$$u\,\boldsymbol{t}\,(v\,\boldsymbol{s_t}\,w) = (u\,\boldsymbol{t}\,v)\,\boldsymbol{s_t}\,(u\,\boldsymbol{t}\,w)$$

holds true. Taking $v = w = 1$ gives

$$u = u\,\boldsymbol{s_t}\,u \quad \text{for all } u \in [0,1],$$

i.e. the idempotency of $\boldsymbol{s_t}$. By corollary 1.9 then $\boldsymbol{t}$ is also idempotent and hence $\boldsymbol{t} = \min$.

(ii) Now the distributivity condition is that for all $u, v, w \in [0,1]$ there holds true

$$u\,\boldsymbol{s_t}\,(v\,\boldsymbol{t}\,w) = (u\,\boldsymbol{s_t}\,v)\,\boldsymbol{t}\,(u\,\boldsymbol{s_t}\,w).$$

This time choosing $v = w = 0$ gives

$$u = u\,\boldsymbol{t}\,u \quad \text{for all } u \in [0,1],$$

hence again the idempotency of $\boldsymbol{t}$ and thus $\boldsymbol{t} = \min$. QED

Distributivity of a t-norm with respect to its t-conorm or of a t-conorm with respect to its t-norm is one aspect of distributivity between generalized conjunction and disjunction connectives only. Other ones relate e.g. different t-norms, or a t-norm with a t-conorm of another t-norm etc. We will not (try to) discuss this topic in full generality but mention only some very special cases in which one of the involved t-norms is $\boldsymbol{t} = \boldsymbol{t}_G = \min$. Remembering the notational convention (1.26), by abuse of language we will also write here $\wedge = \boldsymbol{t}_G = \min$ and $\vee = \boldsymbol{s}_{\boldsymbol{t}_G} = \max$.

Proposition 1.12 *For all $u, v, w \in [0,1]$ and any t-norm $\boldsymbol{t}$ there hold true the distributive laws*

$$\begin{array}{rrcl}
(i) & u\,\boldsymbol{t}\,(v \wedge w) & = & (u\,\boldsymbol{t}\,v) \wedge (u\,\boldsymbol{t}\,w),\\
(ii) & u\,\boldsymbol{t}\,(v \vee w) & = & (u\,\boldsymbol{t}\,v) \vee (u\,\boldsymbol{t}\,w),\\
(iii) & u\,\boldsymbol{s_t}\,(v \wedge w) & = & (u\,\boldsymbol{s_t}\,v) \wedge (u\,\boldsymbol{s_t}\,w),\\
(iv) & u\,\boldsymbol{s_t}\,(v \vee w) & = & (u\,\boldsymbol{s_t}\,v) \vee (u\,\boldsymbol{s_t}\,w)
\end{array}$$

as well as the subdistributive laws

$$\begin{array}{rrcl}
(v) & u \wedge (v\,\boldsymbol{t}\,w) & \geq & (u \wedge v)\,\boldsymbol{t}\,(u \wedge w),\\
(vi) & u \vee (v\,\boldsymbol{t}\,w) & \geq & (u \vee v)\,\boldsymbol{t}\,(u \vee w),\\
(vii) & u \wedge (v\,\boldsymbol{s_t}\,w) & \leq & (u \wedge v)\,\boldsymbol{s_t}\,(u \wedge w),\\
(viii) & u \vee (v\,\boldsymbol{s_t}\,w) & \leq & (u \vee v)\,\boldsymbol{s_t}\,(u \vee w).
\end{array}$$

Proof. Claims (i) – (iv) are direct consequences of the definitions of t-norms and their t-conorms. The claims (v) – (viii) follow by applying the respective distributivity result from (i) – (iv) to the right side of that subdistributivity and then using simple inequalities for t-norms and t-conorms. QED

1.3 t-norm based connectives

As was already mentioned, the t-norms are considered as truth functions of generalized conjunction operators in a suitable many-valued logic, and the

Φ-operators shall be considered as truth functions of generalized implication operators. In the present section we will support this point of view by presenting a series of formulas (both propositional and first order) involving these new, generalized connectives which under their intended interpretations become (logically) valid. These formulas, all of them generalizations of well known logically valid formulas of classical logic, mark the extent to which t-norms and Φ-operators may count as suitable generalizations of the corresponding classical, two-valued truth functions. Nevertheless, the formulas discussed further on are not only used to support this point of view, but, besides this reason, also have their independent value within our later developments and will be referred to from different proofs and calculations there.

Throughout this section, H, G, F (sometimes with indices) are used to denote well formed formulas of our (enriched) language of many-valued logic.

Definition 1.7 *For each t-norm* $\mathbf{t}$ *we denote by* $\wedge_{\mathbf{t}}$ *the conjunction connective of the language of our system* $\mathbf{L}$ *of many-valued logic that has this t-norm* $\mathbf{t}$ *as its truth function, similarly we use in the language of* $\mathbf{L}$ *for each t-norm* $\mathbf{t}$ *a disjunction connective* $\vee_{\mathbf{t}}$ *with truth function* $\mathbf{s}_{\mathbf{t}}$. *And for each t-norm* $\mathbf{t}$ *which fulfills* $\mathsf{LSC}(\mathbf{t})$ *by* $\rightarrow_{\mathbf{t}}$ *we denote that implication connective of the language of our system* $\mathbf{L}$ *of many-valued logic that has the Φ-operator* $\varphi_{\mathbf{t}}$ *as its truth function.*

Hence we enrich the language of the system $\mathbf{L}$ with as many new t-norm based conjunction, disjunction and implication connectives as seem to be appropriate in each case. In practice that will usually mean including one or two t-norms into the considerations and hence adding one or two such t-norm based connectives of each kind to the language of $\mathbf{L}$. Because of (1.11), with these notational conventions one also has for any given t-conorm a disjunction connective with this t-conorm as truth function. Hence if H is e.g. the conjunction $H_1 \wedge_{\mathbf{t}} H_2$, we have

$$[\![H_1 \wedge_{\mathbf{t}} H_2]\!] = \mathbf{t}([\![H_1]\!],[\![H_2]\!]) = [\![H_1]\!]\,\mathbf{t}\,[\![H_2]\!]. \tag{1.23}$$

Correspondingly we have for the semantic interpretation of the implication connective $\rightarrow_{\mathbf{t}}$ the simple fact that

$$[\![H_1 \rightarrow_{\mathbf{t}} H_2]\!] = \varphi_{\mathbf{t}}([\![H_1]\!],[\![H_2]\!]). \tag{1.24}$$

Sometimes we also use a biimplication operator $\leftrightarrow_{\mathbf{t}}$ connected with a t-norm $\mathbf{t}$ and defined by

$$H_1 \leftrightarrow_{\mathbf{t}} H_2 =_{def} (H_1 \rightarrow_{\mathbf{t}} H_2) \wedge_{\mathbf{t}} (H_2 \rightarrow_{\mathbf{t}} H_2). \tag{1.25}$$

As we later on occasionally have to discuss separately the strongest t-norm $t_G = \min$, as well as the ŁUKASIEWICZ t-norm t_L and their corresponding t-conorms, we use as shorthand notations

$$\wedge \ \text{for} \ \wedge_{\min} \qquad \text{and} \qquad \vee \ \text{for} \ \vee_{\min}, \tag{1.26}$$
$$\& \ \text{for} \ \wedge_{t_L} \qquad \text{and} \qquad \uplus \ \text{for} \ \vee_{t_L}, \tag{1.27}$$

and by abuse of language from time to time those symbols are also used for those t-norms and t-conorms themselves.

Proposition 1.13 *For each t-norm* t *which has the property* $\mathsf{LSC}(t)$ *and all* H_1, H_2 *one has*

$$\models H_1 \rightarrow_t H_2 \quad \textit{iff} \quad [\![H_1]\!] \leq [\![H_2]\!], \tag{1.28}$$
$$\models H_1 \leftrightarrow_t H_2 \quad \textit{iff} \quad [\![H_1]\!] = [\![H_2]\!]. \tag{1.29}$$

Proof. The first claim is a restatement of proposition 1.6 (i) using (1.6), the second one a consequence of the first using: $u\,t\,v = 1 \Leftrightarrow u = v = 1$. QED

Therefore all the logically valid implications of our generalized language are inequalities between truth degrees "in disguise", and the logically valid biimplications accordingly represent equalities for truth degrees.

For simplicity of notation we use the arrows $\rightarrow$ and $\leftrightarrow$ here also without subscripts: in that case we assume that a subscript "t_1" has been deleted that refers to any t-norm t_1 with property $\mathsf{LSC}(t_1)$.

Proposition 1.14 *For any t-norm* t *the following expressions are logically valid:*

$$
\begin{array}{ll}
(i) & \models H \wedge_t \top \leftrightarrow H \quad \textit{and} \quad \models H \wedge_t \bot \leftrightarrow \bot, \\
(ii) & \models H \vee_t \top \leftrightarrow \top \quad \textit{and} \quad \models H \vee_t \bot \leftrightarrow H, \\
(iii) & \models H_1 \wedge_t H_2 \leftrightarrow H_2 \wedge_t H_1, \\
(iv) & \models H_1 \vee_t H_2 \leftrightarrow H_2 \vee_t H_1, \\
(v) & \models H_1 \wedge_t (H_2 \wedge_t H_3) \leftrightarrow (H_1 \wedge_t H_2) \wedge_t H_3, \\
(vi) & \models H_1 \vee_t (H_2 \vee_t H_3) \leftrightarrow (H_1 \vee_t H_2) \vee_t H_3.
\end{array}
$$

Proof. Those statements immediately follow from properties (T1) to (T3) of any t-norm and properties (S1) to (S3) of any t-conorm. QED

Therefore the t-norm based conjunctions and disjunctions are already by definition – modulo logical validity – commutative as well as associative operations. As was already seen for t-norms and t-conorms distributivity properties only hold true for special cases. As an immediate consequence of the results of proposition 1.12 we mention the following two propositions without further proof.

Proposition 1.15 *For any t-norm* $\mathbf{t}$ *the following equivalences are logically valid:*

$$
\begin{array}{ll}
(i) & \models H_1 \wedge_{\mathbf{t}} (H_2 \wedge H_3) \leftrightarrow (H_1 \wedge_{\mathbf{t}} H_2) \wedge (H_1 \wedge_{\mathbf{t}} H_3), \\
(ii) & \models H_1 \wedge_{\mathbf{t}} (H_2 \vee H_3) \leftrightarrow (H_1 \wedge_{\mathbf{t}} H_2) \vee (H_1 \wedge_{\mathbf{t}} H_3), \\
(iii) & \models H_1 \vee_{\mathbf{t}} (H_2 \wedge H_3) \leftrightarrow (H_1 \vee_{\mathbf{t}} H_2) \wedge (H_1 \vee_{\mathbf{t}} H_3), \\
(iv) & \models H_1 \vee_{\mathbf{t}} (H_2 \vee H_3) \leftrightarrow (H_1 \vee_{\mathbf{t}} H_2) \vee (H_1 \vee_{\mathbf{t}} H_3).
\end{array}
$$

Proposition 1.16 *For any t-norm* $\mathbf{t}$ *the following implications are logically valid:*

$$
\begin{array}{ll}
(i) & \models (H_1 \wedge H_2) \wedge_{\mathbf{t}} (H_1 \wedge H_3) \rightarrow H_1 \wedge (H_2 \wedge_{\mathbf{t}} H_3), \\
(ii) & \models (H_1 \vee H_2) \wedge_{\mathbf{t}} (H_1 \vee H_3) \rightarrow H_1 \vee (H_2 \wedge_{\mathbf{t}} H_3), \\
(iii) & \models H_1 \wedge (H_2 \vee_{\mathbf{t}} H_3) \rightarrow (H_1 \wedge H_2) \vee_{\mathbf{t}} (H_1 \wedge H_3), \\
(iv) & \models H_1 \vee (H_2 \vee_{\mathbf{t}} H_3) \rightarrow (H_1 \vee H_2) \vee_{\mathbf{t}} (H_1 \vee H_3).
\end{array}
$$

Proposition 1.17 *For any t-norm* $\mathbf{t}$ *with property* $\mathsf{LSC}(\mathbf{t})$ *the following expressions are logically valid:*

$$
\begin{array}{ll}
(i) & \models H \wedge_{\mathbf{t}} (H \rightarrow_{\mathbf{t}} G) \rightarrow G, \\
(ii) & \models G \rightarrow (H \rightarrow_{\mathbf{t}} H \wedge_{\mathbf{t}} G), \\
(iii) & \models H \rightarrow (G \rightarrow_{\mathbf{t}} H), \\
(iv) & \models (H_1 \rightarrow_{\mathbf{t}} H_2) \wedge_{\mathbf{t}} G \rightarrow (H_1 \rightarrow_{\mathbf{t}} H_2 \wedge_{\mathbf{t}} G).
\end{array}
$$

Proof. The claims (i) and (ii) follow immediately from the defining properties (Φ2), (Φ3) of the Φ-operators. (iii) is proposition 1.4 (iv). And to get (iv) it is enough to have

$$(u \,\varphi_{\mathbf{t}}\, v)\, \mathbf{t}\, w \leq u \,\varphi_{\mathbf{t}} (v \,\mathbf{t}\, w) \tag{1.30}$$

for any $u, v, w \in [0,1]$. To see that this last inequality holds true one starts from

$$(u \,\varphi_{\mathbf{t}}\, v)\, \mathbf{t}\, w = w\, \mathbf{t} \sup\{s \mid u\, \mathbf{t}\, s \leq v\} = \sup\{w\, \mathbf{t}\, s \mid u\, \mathbf{t}\, s \leq v\}.$$

But with $\boldsymbol{u\,t\,s} \leq v$ one has $\boldsymbol{u\,t}(w\,\boldsymbol{t}\,s) = (\boldsymbol{u\,t\,s})\,\boldsymbol{t}\,w \leq v\,\boldsymbol{t}\,w$ and hence

$$\{w\,\boldsymbol{t}\,s \mid u\,\boldsymbol{t}\,s \leq v\} \subseteq \{z \mid u\,\boldsymbol{t}\,z \leq v\,\boldsymbol{t}\,w\}.$$

Therefore one gets

$$(u\,\varphi_{\boldsymbol{t}}\,v)\,\boldsymbol{t}\,w \leq \sup\{z \mid u\,\boldsymbol{t}\,z \leq v\,\boldsymbol{t}\,w\} = u\,\varphi_{\boldsymbol{t}}\,(v\,\boldsymbol{t}\,w)$$

and thus has (1.30) and hence (iv). QED

The monotonicities of the t-norms and the Φ-operators which were stated in (T2), (Φ1) and proposition 1.4 (i) can directly be rewritten as the implications (which, yet, are classical ones, i.e. formulated in the metalanguage):

$$\begin{array}{llllll}
\text{(a)} & \text{if} & \models H_1 \to H_2 & \text{then} & \models H_1 \wedge_{\boldsymbol{t}} G \to H_2 \wedge_{\boldsymbol{t}} G, \\
\text{(b)} & \text{if} & \models H_1 \to H_2 & \text{then} & \models (G \to_{\boldsymbol{t}} H_1) \to (G \to_{\boldsymbol{t}} H_2), \\
\text{(c)} & \text{if} & \models H_1 \to H_2 & \text{then} & \models (H_2 \to_{\boldsymbol{t}} G) \to (H_1 \to_{\boldsymbol{t}} G).
\end{array}$$

But it is also possible to generalize them to inequalities between truth degrees, that means to logically valid "many-valued" implications.

As in classical logic one has a direct connection between the result of proposition 1.17 (i) and the inference rule of *modus ponens*: because of proposition 1.17 (i) the rule of detachment

$$\frac{H,\ H \to_{\boldsymbol{t}} G}{G} \qquad \text{(modus ponens)}$$

is correct, for, having given well formed formulas H and $H \to_{\boldsymbol{t}} G$ such that $\models H$ and $\models H \to_{\boldsymbol{t}} G$, then by proposition 1.17 (i) immediately $\models G$. Hence any application of the modus ponens as a rule of inference leads from logically valid formulas again to a logically valid formula.

As a side remark let us mention that each one of the metatheoretical implications (a),...,(c) which we have just mentioned also provides the correctness proof for some rule of inference; e.g. because of (b) the rule of inference

$$\frac{H_1 \to_{\boldsymbol{t}} H_2}{(G \to_{\boldsymbol{t}} H_1) \to_{\boldsymbol{t}} (G \to_{\boldsymbol{t}} H_2)}$$

is correct. Furthermore, each logically valid many-valued implication of the form

$$\models H_1 \wedge_{\boldsymbol{t}} \ldots \wedge_{\boldsymbol{t}} H_n \to_{\boldsymbol{t}} H$$

yields the correctness of the rule of inference

$$\frac{H_1, \ldots, H_n}{H}.$$

Therefore, as corollaries of the following propositions we get the correctness of a whole series of rules of inference but will not mention this fact in those future cases separately.

Proposition 1.18 *For any t-norm $\boldsymbol{t}$ with property* $\mathsf{LSC}(\boldsymbol{t})$ *the following expressions are logically valid:*

$$\begin{array}{ll} (i) & \models (H_1 \rightarrow_{\boldsymbol{t}} H_2) \rightarrow (H_1 \wedge_{\boldsymbol{t}} G \rightarrow_{\boldsymbol{t}} H_2 \wedge_{\boldsymbol{t}} G), \\ (ii) & \models (H_1 \rightarrow_{\boldsymbol{t}} H_2) \rightarrow ((G \rightarrow_{\boldsymbol{t}} H_1) \rightarrow_{\boldsymbol{t}} (G \rightarrow_{\boldsymbol{t}} H_2)), \\ (iii) & \models (H_1 \rightarrow_{\boldsymbol{t}} H_2) \rightarrow ((H_2 \rightarrow_{\boldsymbol{t}} G) \rightarrow_{\boldsymbol{t}} (H_1 \rightarrow_{\boldsymbol{t}} G)). \end{array}$$

Proof. (i) By the monotonicity condition (T2) one has for each t-norm $\boldsymbol{t}$

$$u \,\boldsymbol{t}\, z \leq v \Rightarrow (u \,\boldsymbol{t}\, w) \,\boldsymbol{t}\, z = (u \,\boldsymbol{t}\, z) \,\boldsymbol{t}\, w \leq v \,\boldsymbol{t}\, w$$

and therefore using the representation lemma

$$u \,\varphi_{\boldsymbol{t}}\, v \leq (u \,\boldsymbol{t}\, w) \,\varphi_{\boldsymbol{t}}\, (v \,\boldsymbol{t}\, w).$$

But that is precisely the statement in its truth functional form.

(ii) What has to be proven is

$$u \,\varphi_{\boldsymbol{t}}\, v \leq (w \,\varphi_{\boldsymbol{t}}\, u) \,\varphi_{\boldsymbol{t}}\, (w \,\varphi_{\boldsymbol{t}}\, v),$$

that means by the representation lemma

$$\sup\{z \mid u \,\boldsymbol{t}\, z \leq v\} \leq \sup\{x \mid (w \,\varphi_{\boldsymbol{t}}\, u) \,\boldsymbol{t}\, z \leq w \,\varphi_{\boldsymbol{t}}\, v\}.$$

Therefore the main point is to prove the claim that always

$$u \,\boldsymbol{t}\, z \leq v \Rightarrow (w \,\varphi_{\boldsymbol{t}}\, u) \,\boldsymbol{t}\, z \leq w \,\varphi_{\boldsymbol{t}}\, v$$

holds true. Thus assume $u \,\boldsymbol{t}\, z \leq v$. Then using (1.30) one gets

$$(w \,\varphi_{\boldsymbol{t}}\, u) \,\boldsymbol{t}\, z \leq w \,\varphi_{\boldsymbol{t}}\, (u \,\boldsymbol{t}\, z) \leq w \,\varphi_{\boldsymbol{t}}\, v$$

and the proof of (ii) is accomplished.

(iii) In the present case one has to prove that always

$$u \,\varphi_{\boldsymbol{t}}\, v \leq (v \,\varphi_{\boldsymbol{t}}\, w) \,\varphi_{\boldsymbol{t}}\, (u \,\varphi_{\boldsymbol{t}}\, w),$$

and gets as in part (ii) via the representation lemma as a sufficient condition for that inequality to hold true to have

$$u\,\mathbf{t}\,z \leq v \Rightarrow (v\,\varphi_{\mathbf{t}}\,w)\,\mathbf{t}\,z \leq u\,\varphi_{\mathbf{t}}\,w$$

for all $u, v, w, z \in [0,1]$. Thus assume $u\,\mathbf{t}\,z \leq v$. By $\mathsf{LSC}(\mathbf{t})$ and this assumption one has

$$\begin{aligned} (v\,\varphi_{\mathbf{t}}\,w)\,\mathbf{t}\,z &= z\,\mathbf{t}\,\sup\{x \mid v\,\mathbf{t}\,x \leq w\} \\ &= \sup\{z\,\mathbf{t}\,x \mid v\,\mathbf{t}\,x \leq w\} \\ &\leq \sup\{z\,\mathbf{t}\,x \mid u\,\mathbf{t}\,z\,\mathbf{t}\,x \leq w\} \end{aligned}$$

and hence

$$(v\,\varphi_{\mathbf{t}}\,w)\,\mathbf{t}\,z \leq \sup\{y \mid u\,\mathbf{t}\,y \leq w\} = u\,\varphi_{\mathbf{t}}\,w,$$

i.e. the sufficiency condition. QED

Proposition 1.19 *For any t-norm* $\mathbf{t}$ *with property* $\mathsf{LSC}(\mathbf{t})$ *the following expressions are logically valid:*

$$\begin{array}{rl} (i) & \models H_1 \rightarrow_{\mathbf{t}} (H_2 \rightarrow_{\mathbf{t}} G) \leftrightarrow (H_1 \wedge_{\mathbf{t}} H_2 \rightarrow_{\mathbf{t}} G), \\ (ii) & \models (H_1 \rightarrow_{\mathbf{t}} H_2) \wedge_{\mathbf{t}} (H_2 \rightarrow_{\mathbf{t}} H_3) \rightarrow (H_1 \rightarrow_{\mathbf{t}} H_3), \\ (iii) & \models (H_1 \rightarrow_{\mathbf{t}} G_1) \wedge_{\mathbf{t}} (H_2 \rightarrow_{\mathbf{t}} G_2) \rightarrow (H_1 \wedge_{\mathbf{t}} G_1 \rightarrow_{\mathbf{t}} H_2 \wedge_{\mathbf{t}} G_2), \\ (iv) & \models (H_1 \rightarrow_{\mathbf{t}} G_1) \wedge (H_2 \rightarrow_{\mathbf{t}} G_2) \rightarrow (H_1 \wedge G_1 \rightarrow_{\mathbf{t}} H_2 \wedge G_2), \\ (v) & \models (H_1 \rightarrow_{\mathbf{t}} G_1) \wedge (H_2 \rightarrow_{\mathbf{t}} G_2) \rightarrow (H_1 \vee G_1 \rightarrow_{\mathbf{t}} H_2 \vee G_2). \end{array}$$

And in the case that the t-norm $\mathbf{t}$ *subdistributes over its t-conorm in the sense that*

$$u\,\mathbf{t}\,(v\,\mathbf{s_t}\,w) \leq (u\,\mathbf{t}\,v)\mathbf{s_t}(u\,\mathbf{t}\,w)$$

holds true for all $u, v, w \in [0,1]$ *then one also has*

$$(vi) \quad \models (H_1 \rightarrow_{\mathbf{t}} G_1) \wedge_{\mathbf{t}} (H_2 \rightarrow_{\mathbf{t}} G_2) \rightarrow (H_1 \vee_{\mathbf{t}} G_1 \rightarrow_{\mathbf{t}} H_2 \vee_{\mathbf{t}} G_2).$$

Proof. (i) It is enough to prove for all $u, v, w \in [0,1]$ that always

$$u\,\varphi_{\mathbf{t}}\,(v\,\varphi_{\mathbf{t}}\,w) = (u\,\mathbf{t}\,v)\,\varphi_{\mathbf{t}}\,w,$$

which means by the representation lemma and (T1) to prove

$$\sup\{z \mid u\,\mathbf{t}\,z \leq v\,\varphi_{\mathbf{t}}\,w\} = \sup\{z \mid v\,\mathbf{t}\,u\,\mathbf{t}\,z \leq w\}.$$

To get this equality it is sufficient to always have that

$$u\,\mathbf{t}\,z \leq v\,\varphi_{\mathbf{t}}\,w \Leftrightarrow v\,\mathbf{t}\,u\,\mathbf{t}\,z \leq w.$$

But if $u\,\mathbf{t}\,z \leq v\,\varphi_{\mathbf{t}}\,w$ one has by (T2)

$$v\,\mathbf{t}\,(u\,\mathbf{t}\,z) \leq v\,\mathbf{t}\,\sup\{s \mid v\,\mathbf{t}\,s \leq w\} = \sup\{v\,\mathbf{t}\,s \mid v\,\mathbf{t}\,s \leq w\} \leq w;$$

and conversely in the case that $v\,\mathbf{t}\,u\,\mathbf{t}\,z \leq w$ one has by ($\Phi 3$)

$$u\,\mathbf{t}\,z \leq v\,\varphi_{\mathbf{t}}(v\,\mathbf{t}\,(u\,\mathbf{t}\,z)) \leq v\,\varphi_{\mathbf{t}}\,w.$$

(ii) It is enough to prove for all $u, v, w \in [0,1]$ that always

$$(u\,\varphi_{\mathbf{t}}\,v)\,\mathbf{t}\,(v\,\varphi_{\mathbf{t}}\,w) \leq (u\,\varphi_{\mathbf{t}}\,w).$$

But using again the representation lemma one has

$$\begin{aligned}\sup\{z\,\mathbf{t}\,(v\,\varphi_{\mathbf{t}}\,w) \mid u\,\mathbf{t}\,z \leq v\} &= \sup\{z\,\mathbf{t}\,s \mid u\,\mathbf{t}\,z \leq v \text{ and } v\,\mathbf{t}\,s \leq w\}\\ &\leq \sup\{r \mid u\,\mathbf{t}\,r \leq w\}\end{aligned}$$

because one has $u\,\mathbf{t}\,(z\,\mathbf{t}\,s) \leq w$ immediately from $u\,\mathbf{t}\,z \leq v$ and $v\,\mathbf{t}\,s \leq w$. Thus (ii) follows.

(iii) From (1.30) one gets

$$\begin{aligned}(u\,\varphi_{\mathbf{t}}\,v)\,\mathbf{t}\,(w\,\varphi_{\mathbf{t}}\,z) &\leq u\,\varphi_{\mathbf{t}}(v\,\mathbf{t}\,(w\,\varphi_{\mathbf{t}}\,z))\\ &\leq u\,\varphi_{\mathbf{t}}(w\,\mathbf{t}\,(v\,\varphi_{\mathbf{t}}\,z)) = (u\,\mathbf{t}\,w)\,\varphi_{\mathbf{t}}(v\,\mathbf{t}\,z)\end{aligned}$$

as in the proof of (i) and thus also has (iii).

(iv) What we have to prove for any $u_1, u_2, v_1, v_2 \in [0,1]$ is that

$$\min\{u_1\,\varphi_{\mathbf{t}}\,v_1, u_2\,\varphi_{\mathbf{t}}\,v_2\} \leq (\min\{u_1, u_2\})\,\varphi_{\mathbf{t}}\,(\min\{v_1, v_2\}).$$

Without loss of generality we may suppose that $u_1 \leq u_2$. Then we have to prove the simpler inequality

$$\min\{u_1\,\varphi_{\mathbf{t}}\,v_1, u_2\,\varphi_{\mathbf{t}}\,v_2\} \leq u_1\,\varphi_{\mathbf{t}}(\min\{v_1, v_2\}). \tag{1.31}$$

But if $v_1 \leq v_2$, the right hand side of (1.31) becomes $u_1\,\varphi_{\mathbf{t}}\,v_1$ and thus is one of the terms over which on the left hand side the minimum has to be taken – hence (1.31) holds true in this case. And in the case that $v_2 < v_1$, then one has $u_1\,\mathbf{t}\,w \leq u_2\,\mathbf{t}\,w$ for any $w \in [0,1]$ and hence $u_2\,\varphi_{\mathbf{t}}\,v_2 \leq u_1\,\varphi_{\mathbf{t}}\,v_1$. Therefore (1.31) becomes

$$u_2\,\varphi_{\mathbf{t}}\,v_2 \leq u_1\,\varphi_{\mathbf{t}}\,v_2$$

which obviously holds true because of $u_1 \leq u_2$.

(v) What we now have to prove for any $u_1, u_2, v_1, v_2 \in [0,1]$ is

$$\min\{u_1 \varphi_t v_1, u_2 \varphi_t v_2\} \leq (\max\{u_1,u_2\}) \varphi_t (\max\{v_1,v_2\}).$$

Without loss of generality we may suppose that $u_1 \varphi_t v_1 \leq u_2 \varphi_t v_2$. If then for some $w \in [0,1]$ one has $u_1 \, t \, w \leq v_1$, so also $u_2 \, t \, w \leq v_2$; because otherwise in case $v_2 < u_2 \, t \, w$ one would have $w > u_2 \varphi_t v_2$ and thus $u_1 \varphi_t v_1 > u_2 \varphi_t v_2$, a contradiction.

Therefore we now have $\min\{u_1 \varphi_t v_1, u_2 \varphi_t v_2\} = u_1 \varphi_t v_1$ and furthermore

$$\begin{aligned} u_1 \varphi_t v_1 &= \sup\{w \mid u_1 \, t \, w \leq v_1\} \\ &\leq \sup\{w \mid \max\{u_1 \, t \, w, u_2 \, t \, w\} \leq \max\{v_1, v_2\}\} \\ &= \sup\{w \mid \max\{u_1, u_2\} \, t \, w \leq \max\{v_1, v_2\}\} \\ &= (\max\{u_1,u_2\}) \varphi_t (\max\{v_1,v_2\}), \end{aligned}$$

hence all is revealed.

(vi) This time we have to prove for any $u_1, u_2, v_1, v_2 \in [0,1]$ that

$$(u_1 \varphi_t v_1) \, t \, (u_2 \varphi_t v_2) \leq (u_1 \, s_t \, u_2) \varphi_t (v_1 \, s_t \, v_2).$$

Because of $\mathsf{LSC}(t)$ we first have

$$\begin{aligned} &(u_1 \varphi_t v_1) \, t \, (u_2 \varphi_t v_2) \\ &\quad = \sup\{w \mid u_1 \, t \, w \leq v_1\} \, t \, \sup\{z \mid u_2 \, t \, z \leq v_2\} \\ &\quad = \sup\{w \, t \, \sup\{z \mid u_2 \, t \, z \leq v_2\} \mid u_1 \, t \, w \leq v_1\} \\ &\quad = \sup\{\sup\{w \, t \, z \mid u_2 \, t \, z \leq v_2\} \mid u_1 \, t \, w \leq v_1\} \\ &\quad = \sup\{w \, t \, z \mid u_1 \, t \, w \leq v_1 \text{ and } u_2 \, t \, z \leq v_2\} \\ &\quad \leq \sup\{w \, t \, z \mid (u_1 \, t \, w) \, s_t \, (u_2 \, t \, z) \leq v_1 \, s_t \, v_2\}. \end{aligned}$$

By our subdistributivity assumption we now get

$$(u_1 \, t \, w) \, s_t \, (u_2 \, t \, z) \geq (u_1 \, s_t \, u_2) \, t \, (w \, t \, z)$$

for all $w, z \in [0,1]$, for we may, by symmetry, additionally suppose that $w \geq z$ and then have the following sequence of inequalities

$$\begin{aligned} (u_1 \, t \, w) \, s_t \, (u_2 \, t \, z) &\geq (u_1 \, t \, z) \, s_t \, (u_2 \, t \, z) \\ &\geq (u_1 \, s_t \, u_2) \, t \, z \\ &\geq (u_1 \, s_t \, u_2) \, t \, (w \, t \, z). \end{aligned}$$

Thus we have

$$\begin{aligned}(u_1 \, \mathbf{s_t} \, u_2) \, \varphi_{\mathbf{t}} \, (v_1 \, \mathbf{s_t} \, v_2) &= \sup\{q \mid (u_1 \, \mathbf{s_t} \, u_2) \, \mathbf{t} \, q \leq v_1 \, \mathbf{s_t} \, v_2\} \\ &\geq \sup\{w \, \mathbf{t} \, z \mid (u_1 \, \mathbf{s_t} \, u_2) \, \mathbf{t} \, (w \, \mathbf{t} \, z) \leq v_1 \, \mathbf{s_t} \, v_2\} \\ &\geq \sup\{w \, \mathbf{t} \, z \mid (u_1 \, \mathbf{t} \, w) \, \mathbf{s_t} \, (u_2 \, \mathbf{t} \, z) \leq v_1 \, \mathbf{s_t} \, v_2\}.\end{aligned}$$

All these inequalities together prove the inequality we had to establish in this case. QED

The standard connections (1.9), (1.10) between t-norms and t-conorms can now be restated as general DE MORGAN laws for $\wedge_{\mathbf{t}}$ and $\vee_{\mathbf{t}}$ with respect to the (standard) negation $\neg$.

Proposition 1.20 *For any t-norm* $\mathbf{t}$ *the following expressions are logically valid:*

$$(i) \quad \models \neg(H_1 \wedge_{\mathbf{t}} H_2) \leftrightarrow (\neg H_1 \vee_{\mathbf{t}} \neg H_2),$$
$$(ii) \quad \models \neg(H_1 \vee_{\mathbf{t}} H_2) \leftrightarrow (\neg H_1 \wedge_{\mathbf{t}} \neg H_2).$$

Proof. Obvious.

The negation functions $\mathbf{n_t}$ which were introduced in definition 1.5 can now also be used to consider t-norm based negations.

Definition 1.8 *For any t-norm* $\mathbf{t}$ *which has the property* LSC($\mathbf{t}$) *let* $-_{\mathbf{t}}$ *be that (unary)* negation *connective which has* $\mathbf{n_t}$ *as its truth function, that means always put*

$$[\![-_{\mathbf{t}} H]\!] = \mathbf{n_t}([\![H]\!]).$$

Proposition 1.21 *For any t-norm* $\mathbf{t}$ *with property* LSC($\mathbf{t}$) *and all formulas* H *it holds true*

$$\models -_{\mathbf{t}} H \leftrightarrow (H \rightarrow_{\mathbf{t}} \bot).$$

Proof. Obvious by comparing the respective truth functions. QED

This way of connecting a negation operator with an implication operator was for example used by K. GÖDEL (1932) in his studies on intuitionistic logic, and was also present in the work of the ŁUKASIEWICZ group in their development of many-valued logic; cf. ŁUKASIEWICZ/TARSKI (1930).

It is interesting to note that for the ŁUKASIEWICZ conjunction & with truth function $\boldsymbol{t}_L$ the negation $-_{\boldsymbol{t}_L}$ is nothing other than the standard negation $\neg$. And furthermore in that special case one also has

$$\models (H \rightarrow_{\boldsymbol{t}_L} G) \Leftrightarrow (\neg H \vee_{\boldsymbol{t}_L} G).$$

Proposition 1.22 *For any t-norm* $\boldsymbol{t}$ *with the property* $\mathsf{LSC}(\boldsymbol{t})$ *the following expressions are logically valid:*

$$(i) \quad \models H \rightarrow -_{\boldsymbol{t}} -_{\boldsymbol{t}} H,$$
$$(ii) \quad \models -_{\boldsymbol{t}} H \leftrightarrow -_{\boldsymbol{t}} -_{\boldsymbol{t}} -_{\boldsymbol{t}} H.$$

Proof. (i) is an immediate corollary of the representation lemma and the property $\mathsf{LSC}(\boldsymbol{t})$.

(ii) Because of (i) one only has to show that

$$((u \varphi_{\boldsymbol{t}} 0) \varphi_{\boldsymbol{t}} 0) \varphi_{\boldsymbol{t}} 0 \leq u \varphi_{\boldsymbol{t}} 0 \tag{1.32}$$

for all $u \in [0,1]$. By the representation lemma and $\mathsf{LSC}(\boldsymbol{t})$ one has

$$\begin{aligned} ((u \varphi_{\boldsymbol{t}} 0) \varphi_{\boldsymbol{t}} 0) \varphi_{\boldsymbol{t}} 0 &= \sup\{w \mid \sup\{v \mid (u \varphi_{\boldsymbol{t}} 0)\, \boldsymbol{t}\, v \leq 0\}\, \boldsymbol{t}\, w \leq 0\} \\ &= \sup\{w \mid \sup\{v\, \boldsymbol{t}\, w \mid (u \varphi_{\boldsymbol{t}} 0)\, \boldsymbol{t}\, v \leq 0\} \leq 0\}. \end{aligned}$$

Now suppose that $w_0 \in [0,1]$ is one of the values which belong to the set whose supremum has to be considered here, i.e. is chosen such that

$$\sup\{v\, \boldsymbol{t}\, w_0 \mid (u \varphi_{\boldsymbol{t}} 0)\, \boldsymbol{t}\, v \leq 0\} \leq 0.$$

Then for each $v \in [0,1]$

$$\sup\{z\, \boldsymbol{t}\, v \mid u\, \boldsymbol{t}\, z \leq 0\} = (u \varphi_{\boldsymbol{t}} 0)\, \boldsymbol{t}\, v \leq 0 \Rightarrow v\, \boldsymbol{t}\, w_0 \leq 0,$$

which means that for all $z, v \in [0,1]$ one has

$$(u\, \boldsymbol{t}\, z \leq 0 \Rightarrow z\, \boldsymbol{t}\, v \leq 0) \Rightarrow v\, \boldsymbol{t}\, w_0 \leq 0. \tag{1.33}$$

Assuming

$$w_0 > u \varphi_{\boldsymbol{t}} 0 = \sup\{z \mid u\, \boldsymbol{t}\, z \leq 0\}$$

then gives $w_0 > 0$ and $u\, \boldsymbol{t}\, w_0 > 0$, i.e. $(u\, \boldsymbol{t}\, w_0 \leq 0 \Rightarrow w_0\, \boldsymbol{t}\, v \leq 0)$ holds true for any v and hence $v\, \boldsymbol{t}\, w_0 \leq 0$ for each v by (1.33). For $v = 1$ this gives $w_0 = 1\, \boldsymbol{t}\, w_0 \leq 0$ and thus a contradiction. Hence $w_0 \leq u \varphi_{\boldsymbol{t}} 0$ and thus (1.32) by the choice of w_0. QED

Proposition 1.23 *For any t-norm* $\mathbf{t}$ *with the property* LSC($\mathbf{t}$) *the following expressions are logically valid:*

$$\begin{array}{ll} (i) & \models -_{\mathbf{t}}(H \wedge_{\mathbf{t}} -_{\mathbf{t}}H), \\ (ii) & \models H \rightarrow_{\mathbf{t}} (-_{\mathbf{t}}H \rightarrow_{\mathbf{t}} G), \\ (iii) & \models (H \rightarrow_{\mathbf{t}} -_{\mathbf{t}}G) \leftrightarrow (G \rightarrow_{\mathbf{t}} -_{\mathbf{t}}H), \\ (iv) & \models (H \rightarrow_{\mathbf{t}} G) \rightarrow (-_{\mathbf{t}}G \rightarrow_{\mathbf{t}} -_{\mathbf{t}}H), \\ (v) & \models (H \rightarrow_{\mathbf{t}} G) \wedge_{\mathbf{t}} (H \rightarrow_{\mathbf{t}} -_{\mathbf{t}}G) \rightarrow -_{\mathbf{t}}(H \wedge_{\mathbf{t}} H). \end{array}$$

Proof. (i) immediately follows from $u\,\mathbf{t}\,(u\,\varphi_{\mathbf{t}}\,0) \leq 0$, which is an instance of ($\Phi 2$). (ii) follows from (i) using proposition 1.19 (i) together with $0\,\varphi_{\mathbf{t}}\,v = 1$ for any v.

(iii) can be split into two problems, the first one to prove

$$\models (H \rightarrow_{\mathbf{t}} -_{\mathbf{t}}G) \rightarrow (G \rightarrow_{\mathbf{t}} -_{\mathbf{t}}H),$$

or equivalently to prove

$$\models (H \rightarrow_{\mathbf{t}} -_{\mathbf{t}}G) \rightarrow_{\mathbf{t}} (G \rightarrow_{\mathbf{t}} -_{\mathbf{t}}H), \tag{1.34}$$

and the second one to prove that the reverse implication from (1.34) is valid too. Because of the symmetry of both problems it is enough only to prove (1.34). Using proposition 1.19 (i) the problem (1.34) becomes to prove

$$\models (H \rightarrow_{\mathbf{t}} -_{\mathbf{t}}G) \wedge_{\mathbf{t}} G \rightarrow_{\mathbf{t}} -_{\mathbf{t}}H, \tag{1.35}$$

but with proposition 1.17 (iv) one has

$$[\![(H \rightarrow_{\mathbf{t}} -_{\mathbf{t}}G) \wedge_{\mathbf{t}} G]\!] \leq [\![H \rightarrow_{\mathbf{t}} -_{\mathbf{t}}G \wedge_{\mathbf{t}} G]\!] \leq [\![H \rightarrow_{\mathbf{t}} \bot]\!] = [\![-_{\mathbf{t}}H]\!]$$

from definition 1.8, which is (1.35). Hence (1.34) is proven.

(iv) can equivalently be stated in the form (1.34) but with G and $-_{\mathbf{t}}G$ exchanged. Thus the same calculations as in (iii) yield the proof.

Finally (v) is an easy consequence of proposition 1.19 (iii) using (i), i.e. using $[\![G \wedge_{\mathbf{t}} -_{\mathbf{t}}G]\!] = 0$. QED

Corollary 1.24 *For the t-norm* $\mathbf{t} = \mathbf{t}_L$ *contraposition holds true in the more traditional form*

$$\models (H_1 \rightarrow_{\mathbf{t}_L} H_2) \leftrightarrow (\neg H_2 \rightarrow_{\mathbf{t}_L} \neg H_1),$$

and also for each t-norm $\boldsymbol{t}$ *with the property* $\mathsf{LSC}(\boldsymbol{t})$, *for which additionally there holds true* $\models (-_{\boldsymbol{t}} -_{\boldsymbol{t}} H \leftrightarrow H)$, *even*

$$\models (H_1 \rightarrow_{\boldsymbol{t}} H_2) \leftrightarrow (-_{\boldsymbol{t}} H_2 \rightarrow_{\boldsymbol{t}} -_{\boldsymbol{t}} H_1)$$

holds true.

Proof. Because of $\models (\neg\neg H \leftrightarrow H)$ both cases can be proven by the same argument: simply substitute $\neg G$ resp. $-_{\boldsymbol{t}} G$ for G in proposition 1.23 (iii) and then delete the double negations. QED

The foregoing discussions of the present section were all concerned with propositional many-valued logic. That is by far the most important topic for t-norm based logical operators. One could also try, of course, to extend these t-norm-based considerations to the realm of quantifiers and to define new t-norm based universal quantifiers which for interpretations with finite universes reduce to iterations of a t-norm-based conjunction connective $\wedge_{\boldsymbol{t}}$, as well as to extend them to new t-norm-based existential quantifiers, which under the same restrictions concerning the interpretations reduce to an iteration of a t-conorm-based disjunction connective. But at present it seems that such a generalization strategy would not produce interesting quantifiers. One of the main difficulties, it seems, comes from the fact that all the t-norms besides $\boldsymbol{t} = \min$ are not idempotent and that therefore the quantifiers based on such non-idempotent t-norms behave quite differently in different interpretations depending on the cardinalities of the respective universes.

In any case, finite iterations of these t-norm-based connectives, mainly conjunction connectives $\wedge_{\boldsymbol{t}}$ and disjunction connectives $\vee_{\boldsymbol{t}}$, can be used in the usual way through inductive generalizations. Only in a few situations later will on the need arise for infinitary generalizations of such propositional connectives. In those cases another way round shall be taken instead of introducing new quantifiers: infinitely long expressions (with a definite length) will then be used. This method proves to be a suitable substitute to use instead of giving the above mentioned generalization of the quantifiers.

Thus we restrict the considerations to those cases which combine the usual quantifiers of many-valued first order logic as defined in (1.1), (1.2), and in (1.3), (1.4) in their bounded form, with t-norm based propositional connectives.

For the Łukasiewicz negation $-_{\boldsymbol{t}_L} = \neg$ one immedately has generalizations of the well known DE MORGAN laws connecting those quantifiers in classical logic.

Proposition 1.25 *The following expressions are logically valid:*

$$\begin{array}{ll} (i) & \models \neg\forall x H(x) \leftrightarrow \exists x \neg H(x), \\ (ii) & \models \neg\exists x H(x) \leftrightarrow \forall x \neg H(x), \\ (iii) & \models \forall x H(x) \leftrightarrow \neg\exists x \neg H(x), \\ (iv) & \models \exists x H(x) \leftrightarrow \neg\forall x \neg H(x). \end{array}$$

Proof. Straightforward from the corresponding definitions.

Proposition 1.26 *For any t-norm* $\mathbf{t}$ *the following expressions are logically valid:*

$$\begin{array}{ll} (i) & \models \forall x H_1(x) \wedge_{\mathbf{t}} \forall x H_2(x) \rightarrow \forall x (H_1(x) \wedge_{\mathbf{t}} H_2(x)), \\ (ii) & \models \exists x (H_1(x) \wedge_{\mathbf{t}} H_2(x)) \rightarrow \exists x H_1(x) \wedge_{\mathbf{t}} \exists x H_2(x), \\ (iii) & \models \forall x H_1(x) \vee_{\mathbf{t}} \forall x H_2(x) \rightarrow \forall x (H_1(x) \vee_{\mathbf{t}} H_2(x)), \\ (iv) & \models \exists x (H_1(x) \vee_{\mathbf{t}} H_2(x)) \rightarrow \exists x H_1(x) \vee_{\mathbf{t}} \exists x H_2(x). \end{array}$$

Proof. (i) For each $a \in \mathcal{X}$ one has $[\![\forall x H_i(x)]\!] \leq [\![H_i(a)]\!]$ for $i = 1,2$ and thus

$$[\![\forall x H_1(x) \wedge_{\mathbf{t}} \forall x H_2(x)]\!] \leq [\![H_1(a) \wedge_{\mathbf{t}} H_2(a)]\!].$$

Hence for any variable "y" which does not appear (either at all or at least not free) in "$\forall x H_1(x) \wedge_{\mathbf{t}} \forall x H_2(x)$" one has

$$[\![\forall x H_1(x) \wedge_{\mathbf{t}} \forall x H_2(x)]\!] \leq \inf_{y \in \mathcal{X}} [\![H_1(y) \wedge_{\mathbf{t}} H_2(y)]\!].$$

And this is exactly (i).

(ii) is proven along the same lines but starting from the inequalities $[\![H_i(a)]\!] \leq [\![\exists x H_i(x)]\!]$.

(iii) as well as (iv) follow via corollary 1.24 from (ii) and (i) respectively, and the propositions 1.25 and 1.20 by simple, direct transformations. QED

Proposition 1.27 *For any t-norm* $\mathbf{t}$ *with property* $\mathsf{LSC}(\mathbf{t})$ *the following expressions are logically valid, if additionally* G *does not contain the variable* x *free:*

$$\begin{array}{ll} (i) & \models \forall x (G \rightarrow_{\mathbf{t}} H(x)) \leftrightarrow (G \rightarrow_{\mathbf{t}} \forall x H(x)), \\ (ii) & \models \forall x (H(x) \rightarrow_{\mathbf{t}} G) \leftrightarrow (\exists x H(x) \rightarrow_{\mathbf{t}} G), \\ (iii) & \models \exists x (G \rightarrow_{\mathbf{t}} H(x)) \leftrightarrow (G \rightarrow_{\mathbf{t}} \exists x H(x)), \\ (iv) & \models \exists x (H(x) \rightarrow_{\mathbf{t}} G) \leftrightarrow (\forall x H(x) \rightarrow_{\mathbf{t}} G). \end{array}$$

Proof. We prove only (i) and (ii) because the rest can be proven with exactly the same methods.

(i) To get the needed equality

$$[\![\forall x(G \rightarrow_t H(x))]\!] = [\![G \rightarrow_t \forall x H(x)]\!]$$

for the truth degrees we show two corresponding inequalities. By the definition (1.1) of $[\![\forall \ldots]\!]$ as an infimum one has

$$[\![\forall x(G \rightarrow_t H(x))]\!] \leq [\![G \rightarrow_t H(a)]\!]$$

for each $a \in \mathcal{X}$ and thus

$$\models \forall x(G \rightarrow_t H(x)) \rightarrow_t (G \rightarrow_t H(a)).$$

Using proposition 1.19 (i) that gives

$$\models \forall x(G \rightarrow_t H(x)) \wedge_t G \rightarrow_t H(a)$$

and hence

$$[\![\forall x(G \rightarrow_t H(x)) \wedge_t G]\!] \leq [\![H(a)]\!],$$

which means for any variable "y" which does not appear in the well-formed formula "$\forall x(G \rightarrow_t H(x)) \wedge_t G$":

$$[\![\forall x(G \rightarrow_t H(x)) \wedge_t G]\!] \leq \inf_{y \in \mathcal{X}} [\![H(y)]\!] = [\![\forall x H(x)]\!].$$

Therefore

$$\models \forall x(G \rightarrow_t H(x)) \wedge_t G \rightarrow_t \forall x H(x)$$

and again by proposition 1.19 (i)

$$\models \forall x(G \rightarrow_t H(x)) \rightarrow_t (G \rightarrow_t \forall x H(x)).$$

But that means just

$$[\![\forall x(G \rightarrow_t H(x))]\!] \leq [\![G \rightarrow_t \forall x H(x)]\!].$$

To get the reverse inequality one may start from the fact that one has

$$[\![G \rightarrow_t \forall x H(x)]\!] \leq [\![G \rightarrow_t H(a)]\!]$$

for each $a \in \mathcal{X}$ and thus

$$[\![G \rightarrow_t \forall x H(x)]\!] \leq \inf_{y \in \mathcal{X}} [\![G \rightarrow_t H(y)]\!].$$

But this is already the inequality that is needed.

(ii) As in the proof of claim (i) one has to prove the equation

$$[\![\forall x(H(x) \rightarrow_{\mathbf{t}} G)]\!] = [\![\exists x H(x) \rightarrow_{\mathbf{t}} G]\!].$$

The fact that $[\![\exists x H(x)]\!] \geq [\![H(a)]\!]$ for each $a \in \mathcal{X}$ and that therefore, because of proposition 1.4 (i), $[\![\exists x H(x) \rightarrow_{\mathbf{t}} G]\!] \leq [\![H(a) \rightarrow_{\mathbf{t}} G]\!]$ holds true immediately leads to

$$[\![\exists x H(x) \rightarrow_{\mathbf{t}} G]\!] \leq \inf_{y \in \mathcal{X}} [\![H(y) \rightarrow_{\mathbf{t}} G]\!] = [\![\forall x(H(x) \rightarrow_{\mathbf{t}} G)]\!].$$

Thus the first inequality is proven.

For the other one we use the notation $g = [\![G]\!]$ and $h(x) = [\![H(x)]\!]$ and have via representation lemma to prove that

$$\inf_{x \in \mathcal{X}} (h(x) \varphi_{\mathbf{t}} g) \leq (\sup_{x \in \mathcal{X}} h(x)) \varphi_{\mathbf{t}} g = \sup\{w \mid (\sup_{x \in \mathcal{X}} h(x)) \, \mathbf{t} \, w \leq g\}.$$

But this results from the fact that the leftmost term itself is one of the values over which the supremum has to be taken in the rightmost term, for one has:

$$\begin{aligned}(\sup_{x \in \mathcal{X}} h(x)) \, \mathbf{t} \, \inf_{x \in \mathcal{X}} (h(x) \varphi_{\mathbf{t}} g) &\leq \inf_{x \in \mathcal{X}} ((\sup_{y \in \mathcal{X}} h(y)) \, \mathbf{t} \, (h(x) \varphi_{\mathbf{t}} g)) \\ &= \inf_{x \in \mathcal{X}} \sup_{y \in \mathcal{X}} (h(y) \, \mathbf{t} \, (h(x) \varphi_{\mathbf{t}} g)) \leq \sup_{y \in \mathcal{X}} (h(y) \, \mathbf{t} \, (h(y) \varphi_{\mathbf{t}} g)) \leq g.\end{aligned}$$

Hence the second inequality is proven too. QED

Proposition 1.28 *For any t-norm $\mathbf{t}$ with property $\mathsf{LSC}(\mathbf{t})$ the following expressions are logically valid, if additionally G does not contain the variable x free:*

$$\begin{aligned}&(i) && \models \exists x(G \wedge_{\mathbf{t}} H(x)) \leftrightarrow G \wedge_{\mathbf{t}} \exists x H(x), \\ &(ii) && \models \forall x(H_1(x) \rightarrow_{\mathbf{t}} H_2(x)) \wedge_{\mathbf{t}} \exists x H_1(x) \rightarrow \exists x H_2(x).\end{aligned}$$

Proof. With the notation $g = [\![G]\!]$ and $h(x) = [\![H(x)]\!]$ as in the last proof, for (i) one has to prove

$$\sup_{x \in \mathcal{X}} (g \, \mathbf{t} \, h(x)) = g \, \mathbf{t} \, (\sup_{x \in \mathcal{X}} h(x)),$$

and this is exactly the condition $\mathsf{LSC}(t)$. Thus (i) holds true simply by assumption. And for (ii) one has

$$\begin{aligned}&[\![\forall x(H_1(x) \rightarrow_t H_2(x)) \wedge_t \exists x H_1(x)]\!]\\ &\quad = [\![\exists x(\forall y(H_1(y) \rightarrow_t H_2(y)) \wedge_t H_1(x))]\!]\\ &\quad \leq [\![\exists x((H_1(x) \rightarrow_t H_2(x)) \wedge_t H_1(x))]\!]\\ &\quad \leq [\![\exists x H_2(x)]\!]\end{aligned}$$

using (i) together with (Φ2), and that is all that has to be shown. QED

The most interesting point with item (i) of the last proposition is, that the condition $\mathsf{LSC}(t)$ means exactly the logical validity of that expression, i.e. means that as in classical logic an existential quantifier can be "moved into" a conjunction (now: t-norm based) in case only one of the conjuncts contains the quantified variable. From the logical perspective it is, besides the definability of a corresponding implication operator, the need for this property which makes the assumption $\mathsf{LSC}(t)$ often necessary.

What has been lacking until now with regard to such logically valid formulas which give equivalences describing possibilities for the distribution of quantifiers over suitable conjunctions and implications is the case of universal quantification of a t-norm based conjunction. Indeed, that needs another type of assumption.

Proposition 1.29 *For any t-norm t with property $\mathsf{USC}(t)$ the following expression is logically valid, if additionally G does not contain the variable x free:*

$$\models \forall x(G \wedge_t H(x)) \leftrightarrow G \wedge_t \forall x H(x).$$

Proof. Straightforward calculations as in the last proofs. QED

Therefore the closest analogies with the situation in classical logic are obtained from the assumption of the continuity of the t-norm t which is behind the connectives under consideration. But at present it does not seem to be clear whether this assumption of continuity is too strong, or whether one should accept it in every case. In other words, it is an open problem which discontinuous t-norms (if any) may be of special interest in the realm of fuzzy sets theory. The drastic product t_D is the most well known example of a discontinuous t-norm, but unfortunately (?) it has the property $\mathsf{USC}(t_D)$

and lacks having the property $\mathsf{LSC}(\boldsymbol{t}_D)$, thus there is no Φ-operator, i.e. no implication operator intimately connected to $\boldsymbol{t}_D$.

As the last topic essentially involving the quantifiers, let us reconsider the problem of proposition 1.25 and take a look at the relations between any of the t-norm based negations $-_{\boldsymbol{t}}$ and the quantifiers.

Proposition 1.30 *For any t-norm* $\boldsymbol{t}$ *with property* $\mathsf{LSC}(\boldsymbol{t})$ *the following expressions are logically valid:*

$$(i) \quad \models \exists x -_{\boldsymbol{t}} H(x) \leftrightarrow -_{\boldsymbol{t}}\forall x H(x),$$
$$(ii) \quad \models -_{\boldsymbol{t}}\exists x H(x) \leftrightarrow \forall x -_{\boldsymbol{t}} H(x).$$

Proof. By definition 1.8 and proposition 1.27 (iv) one has

$$[\![\exists x -_{\boldsymbol{t}} H]\!] = [\![\exists x(H \rightarrow_{\boldsymbol{t}} \bot)]\!] = [\![\forall x H \rightarrow_{\boldsymbol{t}} \bot]\!] = [\![-_{\boldsymbol{t}}\forall x H]\!]$$

and hence (i). To get (ii) one has to refer to proposition 1.27 (ii) instead. QED

As a means for the formulation of some later results we also need the following notions and notations.

Definition 1.9 *Suppose* $H_1, H_2, H_3, \ldots$ *is a sequence of well-formed formulas and* Θ *a well-formed formula. Then let be*

$$\prod_{i=1}^{1} H_i =_{def} H_1, \qquad \bigwedge_{i=1}^{1} H_i =_{def} H_1,$$
$$\prod_{i=1}^{n+1} H_i =_{def} (\prod_{i=1}^{n} H_i) \wedge_{\boldsymbol{t}} H_{n+1}, \qquad \bigwedge_{i=1}^{n+1} H_i =_{def} (\bigwedge_{i=1}^{1} H_i) \wedge H_{n+1},$$

for each integer $n \geq 1$ *and additionally*

$$[\Theta]^n =_{def} \prod_{i=1}^{n} \Theta.$$

We have omitted the explicite reference to the t-norm $\boldsymbol{t}$ with our finite iteration $\prod$ of $\wedge_{\boldsymbol{t}}$ and with the exponential notation $[\Theta]^n$. We assume that this causes no difficulties as we do not discuss different t norms simultaneously (besides $\boldsymbol{t} = \min$, which has its own notation).

What we especially need in chapter 3 is the following result concerning the exchangeability of many-valued existential quantification with the exponential notation of that last definition.

Proposition 1.31 *Suppose that the t-norm* $\mathbf{t}$ *has property* $\mathsf{LSC}(\mathbf{t})$. *For each integer* $n \geq 1$ *and each well-formed formula* $H(X)$ *of our many-valued language for fuzzy sets theory there holds true*

$$[\![\exists X([H(X)]^n)]\!] = [\![[\exists X H(X)]^n]\!].$$

Proof. Obviously one has $[\![H(X)]\!] \leq [\![\exists X H(X)]\!]$ and therefore also the inequality $[\![[H(X)]^n]\!] \leq [\![[\exists X H(X)]^n]\!]$, such that

$$[\![\exists X([H(X)]^n)]\!] \leq [\![[\exists X H(X)]^n]\!].$$

But we also have

$$\begin{aligned} [\![[\exists X H(X)]^n]\!] &= [\![\prod_{i=1}^{n}(\exists X H(X))]\!] \\ &= \sup\{ [\![\prod_{i=1}^{n} H(X_i/A_i)]\!] \mid A_1, \ldots, A_n \in I\!F(\mathcal{X})\} \end{aligned}$$

because of $\mathsf{LSC}(\mathbf{t})$. And because

$$[\![\prod_{i=1}^{n} H(X_i/A_i)]\!] \leq \max_{1 \leq i \leq n} [\![H(X_i/A_i)^n]\!]$$

we now get

$$[\![[\exists X H(X)]^n]\!] \leq \sup \{ [\![[H(X/A)]^n]\!] \mid A \in I\!F(\mathcal{X})\} = [\![\exists X([H(X)]^n)]\!].$$

Thus the proof is complete. QED

Chapter 2

Basic fuzzy set theory

2.1 Set algebra for fuzzy sets

It is an old problem of fuzzy set theory, already stressed in ZADEH (1965), that we do not have really convincing arguments for something like a "right choice" of our connectives, i.e. of our operations describing the union and intersection of fuzzy sets. Of course, the problem in the past has been discussed from different points of view, cf. e.g. BELLMAN/GIERTZ (1973), YAGER (1979), GOTTWALD (1979a) and GILES (1976), (1979), THOLE/ZIMMERMANN/ZYSNO (1979), ZIMMERMANN/ZYSNO (1980), HAMACHER (1978) too. Besides those motivational discussions for the choice of "right" connectives for fuzzy set operations there is another mainstream in the current research work: to accept a broad class of possible candidates for fuzzy set operations, to discuss them all in parallel, and to choose concrete ones for concrete applications. In this sense different families of operations have been discussed by for example YAGER (1980), DOMBI (1982), WEBER (1983), all of which proved to be special cases of the t-norms of SCHWEIZER/SKLAR (1961) which we discussed extensively in section 1.2. Hence, if we do not change the set [0,1] of generalized membership degrees, these t-norms and their duals, the t-conorms (which are sometimes called s-norms too), seem to require attention. As is well known, the t-norms generalize the usual conjunction operator and hence define intersection operations for fuzzy sets; correspondingly the t-conorms generalize the usual disjunction operator and thus define union operations for fuzzy sets. Additionally, in the following we have to consider generalized complementation operations for fuzzy sets, and we will use the Φ-operators as generalizations of the implication operator and hence for defining generalized inclusion relations.

As usual, every fuzzy set A on a given universe of discourse $\mathcal{X}$ is uniquely characterized by its membership function $\mu_A : \mathcal{X} \to [0,1]$ and furthermore will be identified with this membership function. That means that defining a fuzzy set A (on $\mathcal{X}$) is to define a function $\mu_A : \mathcal{X} \to [0,1]$.

The approach which is presented here will use to a great extent the language of (a suitable system of) a many-valued logic $\mathbf{L}$ and, from this perspective, add two more points to a simple treatment of fuzzy set theory:

(i) we use the generalized – i.e. two-place many-valued – membership predicate ε and interpret the membership degrees $\mu_A(x)$ as truth degrees, i.e. we put (cf. (1.5))

$$\mu_A(x) = \text{truth value of } \; x \,\varepsilon\, A = [\![x \,\varepsilon\, A]\!],$$

in such a way making obvious on the formal level the strong and far-reaching analogies of classical and fuzzy set theory;

(ii) we extend these analogies further by introducing a suitable many-valued generalization of the very useful classical abstraction terms $\{x \mid \ldots\}$ for "the set of all x such that ..."; these generalized abstraction terms will prove as useful for the definition of fuzzy sets as the corresponding classical abstraction terms do for the definition of crisp sets.

In this book we will neither consider the problem of axiomatizability or axiomatization of the set of all valid formulae, nor discuss if and how to base fuzzy set theory on some system of axioms. Instead, we prefer a "naive" approach which presupposes that we have – for every given universe of discourse $\mathcal{X}$ – a sufficiently good intuitive understanding of the class of fuzzy subsets of $\mathcal{X}$.

Additionally, this approach shall be restricted to such fuzzy sets which would constitute the fuzzy sets of the first level inside a full "cumulative", set theoretical hierarchy of fuzzy sets based on some given class of urelements. In set theoretical terms that means that our fuzzy sets shall only be fuzzy sets of "urelements". Different approaches toward full cumulative hierarchies of fuzzy sets like those ones of CHAPIN (1974/75), GOTTWALD (1979), ZHANG (1980), or WEIDNER (1981) are not relevant for our present purposes.[1]

To introduce, as announced, the generalized abstraction terms, we now give the following

[1] Of course there are no extra assumptions concerning the nature of the elements of our universes of discourse. Therefore they themselves may be fuzzy sets, but this fact will not become recognized by our formalism.

Definition 2.1 *For each universe of discourse $\mathcal{X}$ and each formula $H(x)$ of our set theoretic language of many-valued logic we denote by $\{x \in \mathcal{X} \parallel H(x)\}$ or simply by $\{x \parallel H(x)\}$ that fuzzy set A on $\mathcal{X}$ whose membership function μ_A is characterized by: $\mu_A(a) = [\![H(a)]\!]$ for each $a \in \mathcal{X}$; i.e. $\{x \in \mathcal{X} \parallel H(x)\}$ has the characteristic property*

$$[\![a \varepsilon \{x \in \mathcal{X} \parallel H(x)\}]\!] = [\![H(x/a)]\!] \quad \textit{for all } a \in \mathcal{X}.$$

Furthermore, we sometimes have to use a simple many-valued translation of usual identity, which is defined as

$$[\![a \doteq b]\!] = \begin{cases} 1, & \text{if } a = b \\ 0 & \text{otherwise} \end{cases} \tag{2.1}$$

and will also be allowed to be used in the case that instead of a, b some terms (of a well defined kind) appear.

Definition 2.2 *For every t-norm $\mathbf{t}$, t-conorm $\mathbf{s_t}$ and negation function $\mathbf{n}$ let for any fuzzy sets A, B be their (t-norm-based) intersection and union, as well as their complement:*

$$\begin{aligned} A \cap_{\mathbf{t}} B &=_{def} \{x \parallel x \varepsilon A \wedge_{\mathbf{t}} x \varepsilon B\}, \\ A \cup_{\mathbf{t}} B &=_{def} \{x \parallel x \varepsilon A \vee_{\mathbf{t}} x \varepsilon B\}, \\ \complement_{\mathbf{n}} A &=_{def} \{x \parallel \sim_{\mathbf{n}} (x \varepsilon A)\}. \end{aligned}$$

For the special negation functions $\mathbf{n_t}$, $\mathbf{t}$ any t-norm with the property $\mathsf{LSC}(\mathbf{t})$, and $\mathbf{n_{t_L}}$, a separate, simpler notation shall be used:

$$\complement_{\mathbf{t}} A =_{def} \{x \parallel \sim_{\mathbf{n_t}} (x \varepsilon A)\} \quad \textit{and} \quad \complement A =_{def} \{x \parallel \neg(x \varepsilon A)\}.$$

Thus, our generalized abstraction terms $\{x \parallel H(x)\}$ allow for definitional formulas which on the formal level are strongly analogous to the formulas which in classical set theory usually describe the crisp notions corresponding to the fuzzy ones introduced here. Furthermore these far-reaching formal analogies with the crisp case may often be observed in the formulations of the results as well as in the proofs — and they are a severe indication of the (at least methodological) usefulness of our approach toward fuzziness through (the language of) many-valued logic.

To occasionally simplify notation we follow the common usage of writing in the case of $\mathbf{t} = \mathbf{t}_G = \min$

$$\cap \text{ for } \cap_{\mathbf{t}_G} \qquad \text{and} \qquad \cup \text{ for } \cup_{\mathbf{t}_G}. \tag{2.2}$$

Also, as usual, by $\mathbb{F}(\mathcal{X})$ the class

$$\mathbb{F}(\mathcal{X}) =_{def} [0,1]^{\mathcal{X}}$$

of all fuzzy subsets of $\mathcal{X}$ will be denoted. Here, $\mathbb{F}(\mathcal{X})$ is the set power (or the direct power in algebraic terms) of the set of generalized membership (i.e.: truth) degrees with respect to the universe of discourse as exponent. It is well known that this point of view can be extended to include the consideration of set algebraic operations for fuzzy sets too. That means, having given any families $(t_i)_{i\in I}$ of t-norms, $(s_{t_j})_{j\in J}$ of t-conorms, and $(n_k)_{k\in K}$ of negation functions, the algebraic structure

$$\mathbf{F}(\mathcal{X}) = \langle \mathbb{F}(\mathcal{X}), (\cup_{t_i})_{i\in I}, (\cap_{t_j})_{j\in J}, (\complement_{n_k})_{k\in K} \rangle \tag{2.3}$$

of fuzzy subsets of $\mathcal{X}$ is the direct power $\Pi^{\mathcal{X}}$ of the structure

$$\Pi =_{def} \langle [0,1], (t_i)_{i\in I}, (s_{t_j})_{j\in J}, (n_k)_{k\in K} \rangle$$

of generalized membership degrees. The same remark holds true for any kind of L-fuzzy sets too, but as we are interested in t-norms we have to restrict ourselves here to the consideration of the set [0,1] of generalized membership degrees. But, obviously, our principal idea to treat fuzzy set theory with the help of (the language of) a suitable many-valued logic works in the same way too for other kinds of structures of generalized membership degrees than those we are actually discussing here.

If one speaks on the structures Π and $\mathbf{F}(\mathcal{X})$ one has to use terms referring to the elements of those structures. In our case those terms for the structure Π are exactly our expressions of the language of many-valued logic which are built up from variables for truth degrees – or expressions "$x \,\varepsilon\, A$" with x a variable for elements of $\mathcal{X}$ and (the letter) A a symbol for elements of $\mathbb{F}(\mathcal{X})$ – with the help of the connectives (i.e. operation symbols for truth functions) $\wedge_{t_i}, \vee_{t_j}, \sim_{n_k}$; and the terms for the structure $\mathbf{F}(\mathcal{X})$ are exactly our expressions of the extended set theoretic language which are built up from variables and constants for fuzzy sets with the help of the corresponding operation symbols $\cap_{t_i}, \cup_{t_j}, \complement_{n_k}$ introduced in definition 2.2.

By a *basic* HORN *formula* we now understand a finite disjunction (in the classical metalanguage!) of at most one term equation and negations of term equations otherwise. Thus simple examples of basic HORN formulas with respect to Π are (with p, q as [propositional] variables for generalized truth degrees, i.e. for elements of [0,1])

$$p \wedge_{t_i} q \;=\; q \wedge_{t_i} p,$$
$$p \wedge_{t_i} q \;\neq\; p \wedge_{t_i} (q \vee_{t_j} \sim_{n_k} q),$$

and corresponding basic HORN formulas for $\mathbf{F}(\mathcal{X})$ are e.g.

$$\begin{aligned} A \cap_{t_i} B &= B \cap_{t_i} A, \\ A \cap_{t_i} B &\neq A \cap_{t_i} (B \cup_{t_j} \mathbf{C}_{n_k} B). \end{aligned}$$

Furthermore, by a HORN *formula* we mean such a formula which is built up from basic HORN formulas (with respect to the same structure) using conjunction, existential and universal quantification (of classical metalanguage).

Referring to a well known theorem in classical model theory, cf. e.g. CHANG/KEISLER (1973), we now are able to state many set algebraic laws for fuzzy sets at once.

Theorem 2.1 *Suppose that H is any* HORN *formula (with respect to $\mathbf{F}(\mathcal{X})$). Then the structure $\mathbf{F}(\mathcal{X})$ is a model for H, i.e. H holds true for fuzzy subsets of $\mathcal{X}$, if the* HORN *formula $\hat{H}$ corresponding to H but referring to the structure Π (i.e. $\hat{H}$ is built up from H by changing $\cap_t$ to $\wedge_t$, $\cup_t$ to $\vee_t$, $\mathbf{C}_n$ to $\sim_n$ and by an appropriate change of variables too) holds true in the structure Π.*

Proof. (cf. CHANG/KEISLER (1973), pp. 326 ff.).

As an immediate consequence of this theorem we get the commutativity and associativity of all unions $\cup_t$ and all intersections $\cap_t$ for fuzzy sets from the corresponding properties of t-conorms and t-norms because these properties can be simply formulated by HORN formulas as e.g.

$$\bigwedge_{A,B} (A \cap_t B = B \cap_t A),$$

$$\bigwedge_{A,B,C} (A \cap_t (B \cap_t C) = (A \cap_t B) \cap_t C)$$

for the intersections.

If we add the natural ordering relation $\leq$ to the structure Π we get in the direct power $\mathbf{F}(\mathcal{X})$ an inclusion relation characterized by

$$A \subset B \quad \text{iff} \quad [\![a \,\varepsilon\, A]\!] \leq [\![a \,\varepsilon\, B]\!] \text{ for all } a \in \mathcal{X} \tag{2.4}$$

which already was defined by ZADEH (1965) and which is a partial ordering in the class $\mathbb{F}(\mathcal{X})$ of all fuzzy subsets of $\mathcal{X}$.

It is interesting to note that our theorem can be extended to include these ordering relations too. For this purpose we have to extend what is meant by a basic HORN formula: this is now not only a disjunction of negated or unnegated term equations but a disjunction of – negated or unnegated – term equations or term inequalities $\tau_1' \leq \tau_2'$ and $\tau_1 \subset \tau_2$ for terms τ_1, τ_2 relating to Π and terms τ_1', τ_2' relating to the structure $\mathbf{F}(\mathcal{X})$ of (2.3). Hence, for example, the transitivity of the ordering relation $\leq$ can be described by a HORN formula and therefore our theorem yields

$$A \subset B \quad \text{and} \quad B \subset C \Rightarrow A \subset C$$

for any fuzzy sets A, B, C.

Unfortunately, as we have to refer in our theorem to the classical metalanguage, the binary relation $\subset$ for fuzzy sets is a two-valued relation, i.e. not itself a graded one. Thus one would only have to compare the fuzzy set with their graded membership relation by a two-valued implication relation. Instead, from the intuitive idea of an unsharp boundary for fuzzy sets it would be more natural to have an "unsharp" inclusion relation too for the fuzzy sets. Fortunately, by using our language of many-valued logic it is quite simple and natural to introduce such a graded, i.e. "fuzzified" or many-valued inclusion relation again in strong formal analogy with the usual definition of the inclusion relation. To do this we need only the possibility of universal quantification in our language with many truth degrees and we need a generalized, i.e. many-valued implication operator. But, according to the results of section 1.3 the Φ-operators φ_t are suitable candidates for such implication operators. Hence we give the

Definition 2.3 *For any fuzzy sets A, B and any t-norm $\mathbf{t}$ with property* LSC($\mathbf{t}$) *let*

$$A \subseteqq_{\mathbf{t}} B \quad =_{def} \quad \forall x (x \,\varepsilon\, A \rightarrow_{\mathbf{t}} x \,\varepsilon\, B),$$
$$A \equiv_{\mathbf{t}} B \quad =_{def} \quad A \subseteqq_{\mathbf{t}} B \wedge_{\mathbf{t}} B \subseteqq_{\mathbf{t}} A.$$

The truth degree $[\![A \subseteqq_{\mathbf{t}} B]\!]$ is a degree of containment of A in B and the truth degree $[\![A \equiv_{\mathbf{t}} B]\!]$ is a degree of equality for the fuzzy sets A, B.

To see how these definitions work let us look at the t-norms $\mathbf{t}_G$ and $\mathbf{t}_L$. For a readable formulation of the following results we use besides the *support*

$$\text{supp}\,(A) = \{x \in \mathcal{X} \mid \mu_A(x) > 0\} = \{x \in \mathcal{X} \mid [\![x \,\varepsilon\, A]\!] \neq 0\}$$

of $A \in \mathbb{F}(\mathcal{X})$ for any fuzzy sets $A, B \in \mathbb{F}(\mathcal{X})$ also the *crisp* sets

$$\{A > B\} \quad =_{def} \quad \{x \in \mathcal{X} \mid [\![x \,\varepsilon\, A]\!] > [\![x \,\varepsilon\, B]\!]\}, \tag{2.5}$$
$$\{A \neq B\} \quad =_{def} \quad \{x \in \mathcal{X} \mid [\![x \,\varepsilon\, A]\!] \neq [\![x \,\varepsilon\, B]\!]\}. \tag{2.6}$$

These sets generalize the support in the sense that

$$\operatorname{supp}(A) = \{A > \emptyset\} = \{A \neq \emptyset\}$$

holds true for each $A \in \mathbb{F}(\mathcal{X})$.

Straightforward calculations give for $\boldsymbol{t} = \boldsymbol{t}_G = \min$ the result[2]

$$[\![A \subseteq_{\boldsymbol{t}_G} B]\!] = \inf_{x \in \{A > B\}} [\![x \,\varepsilon\, B]\!] \tag{2.7}$$

with the corollary

$$\operatorname{supp}(A) \setminus \operatorname{supp}(B) \neq \emptyset \;\Rightarrow\; [\![A \subseteq_{\boldsymbol{t}_G} B]\!] = 0. \tag{2.8}$$

Then immediately one also has by definition 2.3

$$\begin{aligned} [\![A \equiv_{\boldsymbol{t}_G} B]\!] &= \inf_{x \in \{A \neq B\}} \min\{[\![x \,\varepsilon\, A]\!], [\![x \,\varepsilon\, B]\!]\} \\ &= \inf_{x \in \{A \neq B\}} [\![x \,\varepsilon\, A \cap B]\!] \end{aligned} \tag{2.9}$$

now with the corollary

$$\operatorname{supp}(A) \neq \operatorname{supp}(B) \;\Rightarrow\; [\![A \equiv_{\boldsymbol{t}_G} B]\!] = 0 \tag{2.10}$$

which indicates that $\equiv_{\boldsymbol{t}_G}$ is quite a strong fuzzified equality.

The other case $\boldsymbol{t} = \boldsymbol{t}_L$ again by elementary calculations first gives

$$[\![A \subseteq_{\boldsymbol{t}_L} B]\!] = 1 - \sup_{x \in \{A > B\}} ([\![x \,\varepsilon\, A]\!] - [\![x \,\varepsilon\, B]\!]) \tag{2.11}$$

and therefore with the auxiliary notation

$$\Delta(A, B) =_{def} \sup_{x \in \{A > B\}} ([\![x \,\varepsilon\, A]\!] - [\![x \,\varepsilon\, B]\!])$$

quite directly

$$[\![A \equiv_{\boldsymbol{t}_L} B]\!] = \max\{0, 1 - (\Delta(A, B) + \Delta(B, A))\}. \tag{2.12}$$

This time, contrary to the results (2.8) and (2.10), both of the claims

$$[\![A \subseteq_{\boldsymbol{t}_L} B]\!] \neq 0 \quad \text{and} \quad \operatorname{supp}(A) \not\subseteq \operatorname{supp}(B)$$

[2] We consider, as usual, the infimum over the empty set as =1 and the supremum over the empty set as =0.

as well as both of the claims

$$[A \equiv_{t_L} B] \neq 0 \quad \text{and} \quad \text{supp}\,(A) \neq \text{supp}\,(B),$$

can be true at once.

To some extent therefore $\subseteqq_{t_L}$ and $\equiv_{t_L}$ better meet the intuition behind the fuzzified inclusion and equality than $\subseteqq_{t_G}$ and $\equiv_{t_G}$, namely the intuition that "small" deviations from the "true", i.e. complete inclusion or equality do not completely falsify the generalized inclusion or equality. In addition, deviations from $\text{supp}\,(A) \subseteq \text{supp}\,(B)$ and $\text{supp}\,(A) = \text{supp}\,(B)$ should surely count as "small" as long as combined with small differences in the membership degrees over the "critical" regions $\text{supp}\,(A) \setminus \text{supp}\,(B)$ and $\text{supp}\,(A) \,\Delta\, \text{supp}\,(B)$.[3]

As in chapter 1 we prefer to formulate the further results of this chapter in the form of the logical validity of formulas. If it is equivalent to formulate such a result as an inequality or as an equation for truth degrees we, as before, shall write $\rightarrow$ and $\leftrightarrow$ without index instead of $\rightarrow_{t_0}$ and $\leftrightarrow_{t_0}$ in all those cases where t_0 is any t-norm subject only to the restriction to have the property $\mathsf{LSC}(t_0)$. But we extend this notation and allow also for $\subseteqq$ and $\equiv$ without index – with the same understanding that this is shorthand for $\subseteqq_{t_0}$ and $\equiv_{t_0}$ with a t-norm t_0 with property $\mathsf{LSC}(t_0)$.

The second one of the graded relations defined in definition 2.3 is a kind of many-valued identity relation and will be considered later on. First let us note that because of proposition 1.6 (i) obviously

$$A \subset B \;\Leftrightarrow\; \models A \subseteqq_t B \;\Leftrightarrow\; \models A \subseteqq B \tag{2.13}$$

and therefore also

$$A = B \;\Leftrightarrow\; \models A \equiv_t B \;\Leftrightarrow\; \models A \equiv B \tag{2.14}$$

Hence $\subseteqq_t$ is a suitable generalization of $\subset$; a fact which once again supports the point of view that treating fuzziness by (the language of) many-valued logic is a useful doing. And $\equiv_t$ is at least nicely related to the usual identity relation for fuzzy sets.[4] Furthermore, for the many-valued inclusion relations

[3] Δ denotes here the symmetric difference of crisp sets.

[4] In proposition 2.4 and later on we shall see that $\equiv_t$ has many properties which are natural generalizations of typical properties of the usual (two valued) identity relation. Indeed, $\equiv_t$ is an interesting candidate for a generalized identity of many-valued logic. This topic is discussed in more detail from the theoretical perspective of a many-valued first-order logic with identity in GOTTWALD (1989).

$\subseteqq_t$ suitable many-valued versions of the fundamental properties of inclusion for crisp sets hold true too.

Proposition 2.2 *For every t-norm* $\mathbf{t}$ *with property* $\mathsf{LSC}(\mathbf{t})$ *and all fuzzy sets* A, B, C *there hold true*

$$\begin{array}{ll} (i) & \models A \subseteqq_t A, \\ (ii) & \models A \subseteqq_t B \to A \cap_t C \subseteqq_t B \cap_t C, \\ (iii) & \models A \subseteqq_t B \wedge_t B \subseteqq_t C \to_t A \subseteqq_t C, \\ (iv) & \models A \subseteqq_t B \to \mathbf{C}_t B \subseteqq_t \mathbf{C}_t A, \end{array}$$

and if additionally $\mathbf{s}_{t_1}$ *is any t-conorm such that* $\mathbf{t}$ *distributes over* $\mathbf{s}_{t_1}$ *we too have*

$$(v) \quad \models A \subseteqq_t B \to A \cup_{t_1} C \subseteqq_t B \cup_{t_1} C.$$

Proof. (i) is obvious. For (ii) we have to use the corresponding definitions 2.2 and 2.3 and the fact that

$$[\![H_1 \to_t H_2]\!] \leq [\![H_1 \wedge_t G \to_t H_2 \wedge_t G]\!]$$

which was proven in proposition 1.18 (i).

(iii) By definition 2.2 we have for the truth degrees

$$\begin{aligned} &[\![A \subseteqq_t B \wedge_t B \subseteqq_t C]\!] \\ &\quad = [\![\forall x(x \varepsilon A \to_t x \varepsilon B) \wedge_t \forall x(x \varepsilon B \to_t x \varepsilon C)]\!] \\ &\quad \leq [\![\forall x((x \varepsilon A \to_t x \varepsilon B) \wedge_t (x \varepsilon B \to_t x \varepsilon C))]\!] \end{aligned}$$

using the fact that according to proposition 1.26 (i)

$$\models \forall x H_1 \wedge_t \forall x H_2 \to \forall x(H_1 \wedge_t H_2)$$

holds true for all formulas H_1, H_2 and all t-norms with the property $\mathsf{LSC}(\mathbf{t})$. We furthermore have

$$\models (H_1 \to_t H_2) \wedge_t (H_2 \to_t H_3) \to_t (H_1 \to_t H_3)$$

for all formulas H_1, H_2, H_3; cf. proposition 1.19 (ii). Thus we get

$$[\![A \subseteqq_t B \wedge_t B \subseteqq_t C]\!] \leq [\![A \subseteqq_t C]\!]$$

which immediately yields our statement.

(iv) becomes, by the same kind of argument as in (ii) and (iii), a straightforward consequence of

$$[\![H_1 \rightarrow_t H_2]\!] \le [\![-_t H_2 \rightarrow_t -_t H_1]\!]$$

and this is true in the case of LSC(t) because of proposition 1.23 (iv).

To prove (v) we look as in case (ii) for a proof of

$$[\![A \subseteq_t B]\!] \le [\![A \cup_{t_1} C \subseteq_t B \cup_{t_1} C]\!]$$

which is by definition of $\subseteq_t$ easily established if one is able to show

$$[\![x \varepsilon A \rightarrow_t x \varepsilon B]\!] \le [\![x \varepsilon A \vee_{t_1} x \varepsilon C \rightarrow_t x \varepsilon B \vee_{t_1} x \varepsilon C]\!]$$

Hence we look for a condition of t, t_1 which guarantees for all formulas H_1, H_2, G that

$$[\![H_1 \rightarrow_t H_2]\!] \le [\![H_1 \vee_{t_1} G \rightarrow_t H_2 \vee_{t_1} G]\!],$$

and this means looking for such a condition that yields

$$\sup\{s \mid u\, t\, s \le v\} \le \sup\{t \mid (u\, s_{t_1}\, w)\, t\, t \le v\, s_{t_1}\, w\} \tag{2.15}$$

Now, if t distributes over s_{t_1}, i.e. if always

$$(u\, s_{t_1}\, w)\, t\, t = (u\, t\, t)\, s_{t_1}\, (w\, t\, t),$$

we get immediately always

$$u\, t\, s \le v \Rightarrow (u\, s_{t_1}\, w)\, t\, s \le v\, s_{t_1}\, w$$

and hence (2.15). Therefore (v) is established. QED

Remark. The present proof shows for example that the distributivity of t over s_{t_1} is only a sufficient condition for the monotonicity property

$$\models A \subseteq_t B \rightarrow_t A \cup_{t_1} C \subseteq_t B \cup_{t_1} C,$$

a subdistributivity condition like

$$(u\, s_{t_1}\, w)\, t\, v \le (u\, t\, v)\, s_{t_1}\, (w\, t\, v) \tag{2.16}$$

or simply the property that

$$(u\, s_{t_1}\, w)\, t\, v \le u\, s_{t_1}\, (w\, t\, v) \tag{2.17}$$

(for all $u, v, w \in [0,1]$ of course) are sufficient conditions too.

Furthermore, this proof indicates that (iv) can also be generalized. Considering any negation function $\boldsymbol{n}$ and looking for the truth of

$$\models A \subseteqq_t B \rightarrow \complement_t B \subseteqq_t \complement_t A, \tag{2.18}$$

our line of arguments calls for the contraposition condition

$$u \,\varphi_t\, v \leq \boldsymbol{n}(v)\, \varphi_t\, \boldsymbol{n}(u)$$

to hold true as a sufficient condition for (2.18).

Furthermore, it is possible in a standard way to extend the monotonicity results to for example

$$\models A \subseteqq_t B \wedge_t C \subseteqq_t D \rightarrow A \cap_t C \subseteqq_t B \cap_t D$$

for the case of the intersection $\cap_t$. Besides the monotonicity result of proposition P3.4 (ii) one simply has to use the transitivity result of proposition P3.4 (iii).

Surely, as indicated in this remark, the results of proposition 2.2 can be extended in various ways. But this will not be our problem here, because we mainly intend to show how to work in fuzzy set theory with t-norms, t-conorms and Φ-operators, and how to do this in the context of many-valued logic. Therefore we only add two further special monotonicity results which we will later on have to refer to and then look at the generalized identity relations $\equiv_t$ of definition 2.3 and their fundamental properties.

Proposition 2.3 *For every t-norm* $\boldsymbol{t}$ *with property* $\mathsf{LSC}(\boldsymbol{t})$ *and all fuzzy sets* A, B, C *there hold true*

$$(i) \quad \models A \subseteqq_t B \wedge C \subseteqq_t D \rightarrow A \cap C \subseteqq_t B \cap D,$$
$$(ii) \quad \models A \subseteqq_t B \wedge C \subseteqq_t D \rightarrow A \cup C \subseteqq_t B \cup D.$$

Proof. (i) is an immediate consequence of proposition 1.19 (iv) and (ii) one of proposition 1.19 (v). QED

Proposition 2.4 *Suppose* $\mathsf{LSC}(\boldsymbol{t})$*. Then for all fuzzy sets* A, B, C *there hold true*

$$(i) \quad \models A \equiv_t A,$$
$$(ii) \quad \models A \equiv_t B \rightarrow B \equiv_t A,$$
$$(iii) \quad \models A \equiv_t B \wedge_t B \equiv_t C \rightarrow_t A \equiv_t C,$$
$$(iv) \quad \models A \equiv_t B \rightarrow A \cap_t C \equiv_t B \cap_t C,$$
$$(v) \quad \models A \equiv_t B \rightarrow \complement_t A \equiv_t \complement_t B,$$

and if the t-conorm $\mathbf{s}_{\mathbf{t}_1}$ is such that $\mathbf{t}$ distributes over $\mathbf{s}_{\mathbf{t}_1}$ one also has

$$(vi) \qquad \models A \equiv_{\mathbf{t}} B \rightarrow A \cup_{\mathbf{t}_1} C \equiv_{\mathbf{t}} B \cup_{\mathbf{t}_1} C.$$

Proof. (i) and (ii) are obvious.

(iii) is a simple consequence of the transitivity of $\subseteq_{\mathbf{t}}$, cf. proposition 2.2 (ii), the definition 2.3 of $\equiv_{\mathbf{t}}$, and of the fact that in case $\mathsf{LSC}(\boldsymbol{t})$

$$\models (H_1 \rightarrow_{\mathbf{t}} G_1) \wedge_{\mathbf{t}} (H_2 \rightarrow_{\mathbf{t}} G_2) \rightarrow (H_1 \wedge_{\mathbf{t}} H_2 \rightarrow_{\mathbf{t}} G_1 \wedge_{\mathbf{t}} G_2) \tag{2.19}$$

holds true for all formulas H_1, H_2, G_1, G_2; cf. proposition 1.19 (iii).

(iv) by definition of $\equiv_{\mathbf{t}}$ means

$$\begin{aligned} &\models (A \subseteq_{\mathbf{t}} B \wedge_{\mathbf{t}} B \subseteq_{\mathbf{t}} A) \rightarrow \\ &\quad \rightarrow (A \cap_{\mathbf{t}} C \subseteq_{\mathbf{t}} B \cap_{\mathbf{t}} C \wedge_{\mathbf{t}} B \cap_{\mathbf{t}} C \subseteq_{\mathbf{t}} A \cap_{\mathbf{t}} C) \end{aligned}$$

and hence is a consequence of proposition 2.2 (iii) via (2.19).

Finally, (v) and (vi) can be established in exactly the same way from propositions 2.2 (v) and 2.2 (iv) respectively. QED

For the formulation of further results we need some additional notions. At first these are the empty fuzzy set, fuzzy singletons, and the universal fuzzy set, and later on the intersection and the union of a (crisp) family of fuzzy sets.

Definition 2.4 *For any t-norm $\mathbf{t}$ let be the* universal set X *with respect to the universe of discourse $\mathcal{X}$*

$$X =_{def} \{x \in \mathcal{X} \parallel x \doteq x\},$$

and for each truth degree $u \neq 0$ the u-universal set *with respect to $\mathcal{X}$*

$$X^{[u]} =_{def} \{x \in \mathcal{X} \parallel x \doteq x \wedge_{\mathbf{t}} u\}.$$

The empty fuzzy set $\emptyset$ *(with respect to $\mathcal{X}$) is*

$$\emptyset =_{def} \mathbf{C}X.$$

And for each $a \in \mathcal{X}$ and any truth degree u the (fuzzy) u-singleton of a *is the fuzzy set*

$$\langle\!\langle a \rangle\!\rangle_u =_{def} \{x \parallel x \doteq a \wedge_{\mathbf{t}} u\}.$$

By abuse of notation we have, in the definitions of the u-universal set $X^{[u]}$ and the u-singletons, also taken the truth degree u as the constant of our language denoting just this degree. And we have chosen the same symbol for the empty fuzzy set as for the usual, i.e. crisp empty set. This will not cause any problems and has the advantage of simplicity.

De facto we have that $X^{[1]} = X$ and that each fuzzy singleton $\langle\langle a\rangle\rangle_u$ is independent of the t-norm which is used in this definition because obviously always

$$\begin{aligned} [\![b \,\varepsilon\, X^{[u]}]\!] &= u, \\ [\![b \,\varepsilon\, \langle\langle a\rangle\rangle_u]\!] &= \begin{cases} u, & \text{if } b = a \\ 0 & \text{otherwise} \end{cases} \end{aligned}$$

holds true and additionally

$$[\![b \,\varepsilon\, \emptyset]\!] = 0.$$

Furthermore, $\langle\langle a\rangle\rangle_0 = \emptyset$ is always the case. Obviously too it holds true that

$$\begin{aligned} &\models \;\forall x(x \,\varepsilon\, X), \\ &\models \;\forall x \sim_{\boldsymbol{n}} (x \,\varepsilon\, \emptyset) \end{aligned}$$

for any negation function $\boldsymbol{n}$.

Proposition 2.5 *For each fuzzy set A and every t-norm* $\boldsymbol{t}$ *with the property* $\mathsf{LSC}(\boldsymbol{t})$ *there hold true*

$$\begin{aligned} (i) &\quad \models A \equiv_{\boldsymbol{t}} \emptyset \leftrightarrow \forall x -_{\boldsymbol{t}} (x \,\varepsilon\, A), \\ (ii) &\quad \models -_{\boldsymbol{t}}\exists x(x \,\varepsilon\, A) \leftrightarrow A \equiv_{\boldsymbol{t}} \emptyset, \\ (iii) &\quad \models A \equiv_{\boldsymbol{t}} X \leftrightarrow \forall x(x \,\varepsilon\, A). \end{aligned}$$

Proof. (i) is obvious from the corresponding definitions and proposition 1.4 (ii). (ii) follows from (i) and the fact that

$$\models -_{\boldsymbol{t}}\exists x H \leftrightarrow \forall x -_{\boldsymbol{t}} H$$

holds true; cf. proposition 1.30 (ii). Finally, (iii) again is a simple consequence of the corresponding definitions and proposition 1.4 (iii). QED

The definitions of $\emptyset$ and X are quite natural ones from the point of view of set theory. With the same resulting objects we could have chosen a more algebraic approach: if one includes the truth degrees 0 and 1 as nullary operation in the structure Π then also the direct power $\mathbf{F}(\mathcal{X})$ has to be extended with two nullary operations, and these are exactly $\emptyset$ and X. As our theorem 2.1 (on the transfer of the truth of HORN formulas) covers this case too, we can also get a series of results this way, e.g. for all fuzzy sets A

$$\emptyset \subset A \quad \text{and} \quad A \subset X,$$

or in our many-valued language

$$\models \emptyset \subseteq_t A \quad \text{and} \quad \models A \subseteq_t X$$

for every t-norm $\boldsymbol{t}$. From the fact that

$$p \wedge_t (-_t p) = \bot$$

holds true in Π because of (Φ2) we also get in this way immediately

$$A \cap_t \mathbf{C}_t A = \emptyset,$$

as those formulas are (basic) HORN formulas. Thus we have in our many valued language

$$\models A \cap_t \mathbf{C}_t A \equiv_t \emptyset.$$

But, again, more and stronger results are provable than by this reference only to our transfer theorem 2.1.

Proposition 2.6 *For all t-norms $\boldsymbol{t}$ with property $\mathsf{LSC}(\boldsymbol{t})$, t-conorms $\boldsymbol{s}_{t_1}$ and fuzzy sets A, B there hold true*

$$\begin{array}{ll} (i) & \models A \cap_t B \equiv_t X \to A \equiv_t X \wedge B \equiv_t X, \\ (ii) & \models A \equiv_t X \wedge_t B \equiv_t X \to A \cap_t B \equiv_t X, \\ (iii) & \models A \cup_{t_1} B \equiv_t \emptyset \to A \equiv_t \emptyset \wedge B \equiv_t \emptyset, \end{array}$$

and if $\boldsymbol{t}, \boldsymbol{t}_1$ have the weak DE MORGAN *property that Π is a model of*

$$[\![(-_t p) \wedge_t (-_t q)]\!] \leq [\![-_t (p \vee_{t_1} q)]\!]$$

then it also holds true that

$$(iv) \quad \models A \equiv_t \emptyset \wedge_t B \equiv_t \emptyset \to A \cup_{t_1} B \equiv_t \emptyset.$$

Proof. Let us, because of the additional assumption with regard to (iv), start by proving (iii) and (iv). Then (i) and (ii) can be proven analogously without this additional assumption. The details for those cases are then easily supplied and hence omitted.

(iii) We immediately have

$$\begin{aligned}[\![A \cup_{t_1} B \equiv_t \emptyset]\!] &= [\![\forall x -_t (x \varepsilon A \vee_{t_1} x \varepsilon B)]\!] \\ &\leq [\![\forall x -_t (x \varepsilon A)]\!] = [\![A \equiv_t \emptyset]\!]\end{aligned}$$

from monotonicity properties of $\vee_{t_1}$ and $-_t$. Hence by commutativity of $\cup_{t_1}$:

$$[\![A \cup_{t_1} B \equiv_t \emptyset]\!] \leq \min\{[\![A \equiv_t \emptyset]\!], [\![B \equiv_t \emptyset]\!]\},$$

and hence the result which was stated.

Point (iv) can be analogously derived using

$$\begin{aligned}[\![A \equiv_t \emptyset \wedge_t B \equiv_t \emptyset]\!] &= [\![\forall x -_t (x \varepsilon A) \wedge_t \forall x -_t (x \varepsilon B)]\!] \\ &\leq [\![\forall x(-_t(x \varepsilon A) \wedge_t -_t(x \varepsilon B))]\!],\end{aligned}$$

which holds true according to proposition 1.26 (i), and hence we are able to continue this estimation with

$$[\![A \equiv_t \emptyset \wedge_t B \equiv_t \emptyset]\!] \leq [\![\forall x -_t (x \varepsilon A \vee_{t_1} x \varepsilon B)]\!]$$

because of the assumption on t, t_1 stated additionally for (iv); thus

$$[\![A \equiv_t \emptyset \wedge_t B \equiv_t \emptyset]\!] \leq [\![A \cup_{t_1} B \equiv_t \emptyset]\!]$$

and hence our statement. QED

In order to now define the "big", i.e. infinitary unions and intersections of families of fuzzy sets we will discuss a slight extionsion of our many-valued language: we have to introduce infinitary conjunctions and disjunctions. The finite iterations of each one of the conjunction operators $\wedge_t$ and of each one of the disjunction operators $\vee_t$ are routine and will not be discussed here. What is much more problematic is the infinitary version in general: but we avoid almost all difficulties in restricting ourselves to the t-norm min and the t-conorm max only.

Having given any (crisp) index set I and for each $i \in I$ a formula H_i we introduce as new formulas for infinitary conjunctions and disjunctions

$$\bigwedge\!\!\!\bigwedge_{i \in I} H_i \qquad \text{and} \qquad \bigvee\!\!\!\bigvee_{i \in I} H_i.$$

Those new formulas may be used in building more complex ones as we have so far done with the more simple ones. The truth values of those formulas are given as

$$[\![\bigwedge\!\!\!\bigwedge_{i\in I} H_i]\!] =_{def} \inf_{i\in I}[\![H_i]\!],$$
$$[\![\bigvee\!\!\!\bigvee_{i\in I} H_i]\!] =_{def} \sup_{i\in I}[\![H_i]\!].$$

From these definitions it is obvious that $\bigwedge_{i\in I}$ and $\bigvee_{i\in I}$ will act much like the quantifiers $\forall$ and $\exists$. But we will not prove a detailled list of logically valid formulas including such infinitary parts – if in the following we have to use some special results we will prove them when needed.

Definition 2.5 *For each familiy $(A_i)_{i\in I}$ of fuzzy sets let*

$$\bigcap_{i\in I} A_i \;=_{def}\; \{x \parallel \bigwedge\!\!\!\bigwedge_{i\in I}(x \varepsilon A_i)\},$$
$$\bigcup_{i\in I} A_i \;=_{def}\; \{x \parallel \bigvee\!\!\!\bigvee_{i\in I}(x \varepsilon A_i)\}.$$

As usual, the generalized union can also be used in our fuzzy case to describe every fuzzy set as a union of fuzzy singletons:

$$A = \bigcup_{a\in\mathcal{X}} \langle\!\langle a \rangle\!\rangle_{[\![a\,\varepsilon\,A]\!]}. \tag{2.20}$$

Yet this representation is sometimes too complicated because all the fuzzy singletons $\langle\!\langle a \rangle\!\rangle_0$ are inessential. First, as we mentioned immediately after definition 2.4, $\langle\!\langle a \rangle\!\rangle_0 = \emptyset$ always applies and on the other hand the definition 2.5 generally allows one to disregard in a generalized union $\bigcup_{i\in I} A_i$ all those A_k with $A_k = \emptyset$. Hence instead of (2.20) we simply get

$$A = \bigcup_{a\in \mathrm{supp}(A)} \langle\!\langle a \rangle\!\rangle_{[\![a\,\varepsilon\,A]\!]}, \tag{2.21}$$

which is exactly the same as in ZADEH's (1975) notation the representation of a fuzzy set as a sum or integral of singletons, i.e. is the same as each one of the equalities

$$A = \sum_{a\in \mathrm{supp}(A)} \mu_A(a)/a \qquad \text{or} \qquad A = \int_{a\in \mathrm{supp}(A)} \mu_A(a)/a.$$

Proposition 2.7 *For all fuzzy sets $(A_i)_{i\in I}$, B and arbitrary t-norms $\mathbf{t}$ with the property* $\mathsf{LSC}(\mathbf{t})$ *and t-conorms $\mathbf{s}_{\mathbf{t}_1}$ there hold true:*

$$(i) \qquad \models \bigcap_{i\in I} A_i \subseteqq_{\mathbf{t}} A_k,$$

$$(ii) \qquad \models A_k \subseteqq_{\mathbf{t}} \bigcup_{i\in I} A_i,$$

$$(iii) \qquad \models \bigwedge\!\!\!\bigwedge_{i\in I}(B \subseteqq_{\mathbf{t}} A_i) \leftrightarrow B \subseteqq_{\mathbf{t}} \bigcap_{i\in I} A_i,$$

$$(iv) \qquad \models \bigwedge\!\!\!\bigwedge_{i\in I}(A_i \subseteqq_{\mathbf{t}} B) \leftrightarrow \bigcup_{i\in I} A_i \subseteqq_{\mathbf{t}} B,$$

and without restrictions concerning the involved t-norms there hold true:

$$(v) \qquad \models \bigcap_{i\in I} A_i \cap_{\mathbf{t}} B \subseteqq \bigcap_{i\in I}(A_i \cap_{\mathbf{t}} B),$$

$$(vi) \qquad \models \bigcap_{i\in I} A_i \cup_{\mathbf{t}_1} B \subseteqq \bigcap_{i\in I}(A_i \cup_{\mathbf{t}_1} B),$$

$$(vii) \qquad \models \bigcup_{i\in I}(A_i \cap_{\mathbf{t}} B) \subseteqq \bigcup_{i\in I} A_i \cap_{\mathbf{t}} B,$$

$$(viii) \qquad \models \bigcup_{i\in I}(A_i \cup_{\mathbf{t}_1} B) \subseteqq \bigcup_{i\in I} A_i \cup_{\mathbf{t}_1} B.$$

Proof. (i) and (ii) are obvious. For (iii) we observe that

$$\begin{aligned} [\![\bigwedge\!\!\!\bigwedge_{i\in I}(B \subseteqq_{\mathbf{t}} A_i)]\!] &= \inf_{i\in I} \inf_{x\in\mathcal{X}} [\![x \,\varepsilon\, B \rightarrow_{\mathbf{t}} x \,\varepsilon\, A_i]\!] \\ &= \inf_{x\in\mathcal{X}} [\![x \,\varepsilon\, B \rightarrow_{\mathbf{t}} \bigwedge\!\!\!\bigwedge_{i\in I}(x \,\varepsilon\, A_i)]\!] \end{aligned}$$

according to proposition 1.27 (i). This immediately gives our result. (To be precise, we have to mention that the formulation of proposition 1.27 (i) gives the corresponding result for the quantifier $\forall$; but the proof of that result works exactly the same way for the present infinitary conjunction $\bigwedge$ too.)

To prove (iv) via an equation of suitable truth degrees we proceed analogously but with reference to proposition 1.27 (ii).

Furthermore, (v) to (viii) all are simple consequences of the monotonicity of t-norms and t-conorms. The details can be omitted. QED

Proposition 2.8 *For all fuzzy sets $(A_i)_{i\in I}$, B and arbitrary t-norms $\mathbf{t}$ and t-conorms $\mathbf{s}_{\mathbf{t}_1}$ there hold true:*

(*i*) *in the case of* $\mathsf{USC}(\mathbf{t})$

$$\models \bigcap_{i\in I} A_i \cap_{\mathbf{t}} B \equiv \bigcap_{i\in I}(A_i \cap_{\mathbf{t}} B),$$

(*ii*) *in the case of* $\mathsf{USC}(\mathbf{s}_{\mathbf{t}_1})$

$$\models \bigcap_{i\in I} A_i \cup_{\mathbf{t}_1} B \equiv \bigcap_{i\in I}(A_i \cup_{\mathbf{t}_1} B),$$

(*iii*) *in the case of* $\mathsf{LSC}(\mathbf{t})$

$$\models \bigcup_{i\in I} A_i \cap_{\mathbf{t}} B \equiv \bigcup_{i\in I}(A_i \cap_{\mathbf{t}} B),$$

(*iv*) *in the case of* $\mathsf{LSC}(\mathbf{s}_{\mathbf{t}_1})$

$$\models \bigcup_{i\in I} A_i \cup_{\mathbf{t}_1} B \equiv \bigcup_{i\in I}(A_i \cup_{\mathbf{t}_1} B).$$

Proof. By straightforward calculations from the monotonicity of $\mathbf{t}$ and $\mathbf{s}_{\mathbf{t}_1}$ as mentioned in the last proof. The semicontinuity assumptions in every case guarantee the equality of truth degrees of the left and right sides of the $\equiv_{\mathbf{t}}$-equations. QED

Principally with the same elementary steps as for proving these distributivity properties, one is able to prove the general commutativity and associativity of those infinitary intersections and unions of fuzzy sets, as well as their general distributivity with respect to one another. For formulations of those laws in the crisp case we refer to KLAUA (1964) or any other textbook of set theory which tends toward a more formalized presentation of the topic in the language of first order logic; in the fuzzy case here the laws have to be formulated in the same way. We do not give the details which, in some sense, are routine.

Instead we went to the consideration of cartesian products of fuzzy sets. The problem here is – as later on with fuzzy (binary) relations again – that the cartesian product of two fuzzy subsets of $\mathcal{X}$ need not be a fuzzy subset of $\mathcal{X}$: it depends on $\mathcal{X}$ and on whether $\mathcal{X}$ is closed with respect to (usual) ordered pairing.

Fortunately, for the following discussions of fuzzy cartesian products we do not need to have the universe of discourse $\mathcal{X}$ closed under ordered pairing: we can always work with fuzzy subsets of different universes of discourse.

First let us define the fuzzy cartesian products using the usual notion of an ordered pair (a, b) of elements a, b of the universe of discourse $\mathcal{X}$.

Definition 2.6 *For all fuzzy sets A, B and each t-norm $\mathbf{t}$ let*

$$A \times_{\mathbf{t}} B =_{def} \{(x, y) \,\|\, x \,\varepsilon\, A \wedge_{\mathbf{t}} y \,\varepsilon\, B\}.$$

Obviously we have as an extended meaning of $\subseteqq_{\mathbf{t}}$ and $\equiv_{\mathbf{t}}$ for cartesian products:

$$\begin{aligned} &A \times_{\mathbf{t}} B \subseteqq_{\mathbf{t}_1} C \times_{\mathbf{t}} D =_{def} \\ &\qquad \forall x \forall y (x \,\varepsilon\, A \wedge_{\mathbf{t}} y \,\varepsilon\, B \rightarrow_{\mathbf{t}_1} x \,\varepsilon\, C \wedge_{\mathbf{t}} y \,\varepsilon\, D), \qquad (2.22)\\ &A \times_{\mathbf{t}} B \equiv_{\mathbf{t}_1} C \times_{\mathbf{t}} D =_{def} \\ &\qquad (A \times_{\mathbf{t}} B \subseteqq_{\mathbf{t}_1} C \times_{\mathbf{t}} D) \wedge_{\mathbf{t}_1} (C \times_{\mathbf{t}} D \subseteqq_{\mathbf{t}_1} A \times_{\mathbf{t}} B) \qquad (2.23) \end{aligned}$$

for all fuzzy sets A, B, C, D and all t-norms $\mathbf{t}, \mathbf{t}_1$ where, of course, $\mathbf{t}_1$ has the property $\mathsf{LSC}(\mathbf{t})$.

Set algebraic equations involving fuzzy cartesian products easily result from identities for t-norms, t-conorms and negation functions via our transfer theorem 2.1. To illustrate the method of proof we give some simple examples.

Proposition 2.9 *For t-norms $\mathbf{t}, \mathbf{t}_1$, t-conorms $\mathbf{s}_{\mathbf{t}_2}$ and fuzzy sets A, B, C there hold true*

(i) if $\mathbf{t}$ distributes over $\mathbf{t}_1$:

$$\models A \times_{\mathbf{t}} (B \cap_{\mathbf{t}_1} C) \equiv (A \times_{\mathbf{t}} B) \cap_{\mathbf{t}_1} (A \times_{\mathbf{t}} C),$$

(ii) if $\mathbf{t}$ distributes over $\mathbf{s}_{\mathbf{t}_2}$:

$$\models A \times_{\mathbf{t}} (B \cup_{\mathbf{t}_2} C) \equiv (A \times_{\mathbf{t}} B) \cup_{\mathbf{t}_2} (A \times_{\mathbf{t}} C).$$

Proof. We have by definitions and distributivity of $\mathbf{t}$ over $\mathbf{t}_1$

$$\begin{aligned} A \times_{\mathbf{t}} (B \cap_{\mathbf{t}_1} C) &= \{(x, y) \,\|\, x \,\varepsilon\, A \wedge_{\mathbf{t}} (y \,\varepsilon\, B \wedge_{\mathbf{t}_1} y \,\varepsilon\, C)\} \\ &= \{(x, y) \,\|\, (x \,\varepsilon\, A \wedge_{\mathbf{t}} y \,\varepsilon\, B) \wedge_{\mathbf{t}_1} (x \,\varepsilon\, A \wedge_{\mathbf{t}} y \,\varepsilon\, C)\} \\ &= (A \times_{\mathbf{t}} B) \cap_{\mathbf{t}_1} (A \times_{\mathbf{t}} C) \end{aligned}$$

and hence (i). In the same way the proof of (ii) can proceed. QED

We only mention the fact that the fuzzy cartesian product is commutative and associative – via suitable canonical isomorphisms as in the classical case. Further equalities for fuzzy cartesian products are easily provable by the method of the last proof. We will not discuss this topic more extensively. Instead, we look for proofs of some monotonicity properties and some substitutivity results for fuzzy identities.

Proposition 2.10 *Let $\mathbf{t}$ be a t-norm with property* $\mathsf{LSC}(\mathbf{t})$, *A, B, C and all $(A_i)_{i\in I}$ fuzzy sets. There then hold true*

$$\begin{array}{ll} (i) & \models A \mathrel{\widetilde{\subseteq}_{\mathbf{t}}} B \rightarrow A \times_{\mathbf{t}} C \mathrel{\widetilde{\subseteq}_{\mathbf{t}}} B \times_{\mathbf{t}} C, \\ (ii) & \models A \equiv_{\mathbf{t}} B \rightarrow A \times_{\mathbf{t}} C \equiv_{\mathbf{t}} B \times_{\mathbf{t}} C, \\ (iii) & \models A \equiv_{\mathbf{t}} \emptyset \vee B \equiv_{\mathbf{t}} \emptyset \rightarrow A \times_{\mathbf{t}} B \equiv_{\mathbf{t}} \emptyset, \end{array}$$

which can be sharpened for any t-conorm $\mathbf{s}_{\mathbf{t}_1}$ which has the DE MORGAN *property*

$$[\![-_{\mathbf{t}}(u \wedge_{\mathbf{t}} v)]\!] = [\![(-_{\mathbf{t}}u) \vee_{\mathbf{t}_1} (-_{\mathbf{t}}v)]\!]$$

for all truth degrees u, v to:

$$(iv) \quad \models A \times_{\mathbf{t}} B \equiv_{\mathbf{t}} \emptyset \rightarrow A \equiv_{\mathbf{t}} \emptyset \vee_{\mathbf{t}_1} B \equiv_{\mathbf{t}} \emptyset.$$

With respect to the infinitary unions and intersections in each case one has

$$\begin{array}{ll} (v) & \models \bigcup_{i\in I}(A_i \times_{\mathbf{t}} B) \mathrel{\widetilde{\subseteq}} \bigcup_{i\in I} A_i \times_{\mathbf{t}} B, \\ (vi) & \models \bigcap_{i\in I} A_i \times_{\mathbf{t}} B \mathrel{\widetilde{\subseteq}} \bigcap_{i\in I}(A_i \times_{\mathbf{t}} B), \end{array}$$

with, assuming $\mathsf{LSC}(\mathbf{t})$, *the sharpening*

$$(vii) \quad \models \bigcup_{i\in I}(A_i \times_{\mathbf{t}} B) \equiv \bigcup_{i\in I} A_i \times_{\mathbf{t}} B,$$

and with, assuming $\mathsf{USC}(\mathbf{t})$, *the sharpening*

$$(viii) \quad \models \bigcap_{i\in I} A_i \times_{\mathbf{t}} B \equiv \bigcap_{i\in I}(A_i \times_{\mathbf{t}} B).$$

We omit the proof. But all those statements are analogous to earlier ones relating to for example $\cap_{\mathbf{t}}$ instead of $\times_{\mathbf{t}}$. In every case, too, the proofs needed here are analogous to those earlier proofs which we have presented in detail or sketched with hints for the essential steps.

The part of set algebra for fuzzy sets that was presented here will be representative enough to allow the reader to extend it easily by himself in any of the directions he is interested in.

Therefore, we finish our listing of interesting results of set algebra. Only a slight modification of the construction yielding the fuzzy cartesian products will be mentioned. The point is that the ordered pair (x, y) is a very special object among those which can be built up from x, y. Hence we may assume to have given some binary – or also n-ary, but we will not consider this only slightly more general case – function f on the universe of discourse $\mathcal{X}$. Additionally we assume that each one of the values $f(a, b)$ can be described by a term of our language. To extend our abstraction terms to the case of fuzzy sets with objects $f(a, b)$ as "elements" we define for all formulas $H(x, y)$:

$$\{f(x,y) \,\|\, H(x,y)\} =_{def} \{z \,\|\, \exists x \exists y (H(x,y) \wedge_{\mathbf{t}} z \doteq f(x,y))\} \tag{2.24}$$

for each t-norm $\mathbf{t}$. To avoid any inessential complication let us additionally assume that always $f(x, y) \in \mathcal{X}$, i.e. that f is a binary operation in $\mathcal{X}$. Now we can formulate the famous extension principle of ZADEH (1975) for any t-norm in a very simple manner.

Definition 2.7 (Extension Principle) *Suppose that a binary operation $*$ in the universe of discourse $\mathcal{X}$ is given and some t-norm $\mathbf{t}$ chosen. This operation is extended to a binary operation $*_{\mathbf{t}}$ for fuzzy sets from $\mathbb{F}(\mathcal{X})$ by defining*

$$A *_{\mathbf{t}} B =_{def} \{a * b \,\|\, a \,\varepsilon\, A \wedge_{\mathbf{t}} b \,\varepsilon\, B\}$$

for all fuzzy sets $A, B \in \mathbb{F}(\mathcal{X})$.

Corollary 2.11 *Take the operations $*$ and $*_{\mathbf{t}}$ as in definition 2.7 and consider $C = A *_{\mathbf{t}} B$ for $A, B \in \mathbb{F}(\mathcal{X})$. It then holds true*

$$\mu_C(z) = \sup_{\substack{x,y \in \mathcal{X} \\ z = x * y}} \mu_A(x) \,\mathbf{t}\, \mu_B(y) \qquad \textit{for all } z \in \mathcal{X}.$$

Proof. By (2.24) and the definitions of the quantifier $\exists$ and the class term notation for fuzzy sets, the extension principle immediately yields

$$\mu_C(z) = \sup_{x,y \in \mathcal{X}} [\![x \,\varepsilon\, A \wedge_{\mathbf{t}} b \,\varepsilon\, B]\!] \,\mathbf{t}\, [\![z \doteq x * y]\!].$$

Therefore

$$\mu_C(z) = \sup_{\substack{x,y \in \mathcal{X} \\ z = x * y}} [\![x \varepsilon A \wedge_t b \varepsilon B]\!]$$

because always $[\![z \doteq x * y]\!] \in \{0,1\}$ and only those pairs (x,y) need to be considered here, for which $[\![z \doteq x * y]\!] = 1$, i.e. for which $z = x * y$ is the case.[5] Now

$$[\![x \varepsilon A \wedge_t b \varepsilon B]\!] = \mu_A(x) \, t \, \mu_B(y)$$

gives the result. QED

2.2 Fuzzy relations

Intuitively, a fuzzy relation is a fuzzy set of ordered pairs, i.e. such a fuzzy set whose support is a subset of a suitable crisp cartesian product. That means, we intuitively identify relations with binary ones. This will be enough generality for the present purposes. Of course, mathematically it is obvious that in the same way one may consider n-ary fuzzy relations for any $n \geq 2$.

From the formal point of view we have different possibilities to treat fuzzy relations, e.g. we may

- assume that our universe of discourse $\mathcal{X}$ consists of ordered pairs only, such that each fuzzy subset of $\mathcal{X}$ is a fuzzy relation;
- suppose that the universe of discourse $\mathcal{X}$ is closed with respect to ordered pairing, i.e., we may suppose that for $a, b \in \mathcal{X}$ we have $(a,b) \in \mathcal{X}$ too;
- consider parallel with the universe of discourse $\mathcal{X}$ the set $\mathcal{X} \times \mathcal{X}$ as a second universe and distinguish fuzzy sets as fuzzy subsets of $\mathcal{X}$ from fuzzy relations as fuzzy subsets of $\mathcal{X} \times \mathcal{X}$.

We will essentially follow this third approach here. The first one would allow for a too restricted treatment of fuzzy relations only; and the second approach in principle would force us to take as $\mathcal{X}$ a whole universe for classical set theory (or "nearly" such a whole universe): a situation which we are interested in avoiding here.

---

[5] It is essential for this argumentation that as usual the supremum over the empty set is taken to be =1.

We have to extend our language to treat fuzzy relations. First we add a third sort of symbols: those for fuzzy relations. We use the upper case latin letters

$$R, S, T \qquad \text{(possibly with subscripts or primes)}$$

as symbols for fuzzy relations. As in this chapter we do not have to quantify over fuzzy relations, these symbols will act as constants for fuzzy relations. Of course, as before we will avoid using the letters R, S, T as symbols of fuzzy sets. But, again as before in connection with cartesian products of fuzzy sets, we extend in a straightforward way our use of generalized – i.e. many-valued – abstraction terms to the case of fuzzy relations too. And the same will be done with the notation for the set algebraic operations; i.e. we now take for example

$$R \cap_{\mathbf{t}} S =_{def} \{(x,y) \,\|\, (x,y)\,\varepsilon\, R \wedge_{\mathbf{t}} (x,y)\,\varepsilon\, S\}$$

and in the same manner put

$$R \subseteqq_{\mathbf{t}} S =_{def} \forall x \forall y((x,y)\,\varepsilon\, R \rightarrow_{\mathbf{t}} (x,y)\,\varepsilon\, S).$$

It is straightforward to see that all the results on inclusion and intersection of fuzzy sets, formulated and proved in the last section, also hold true (with slightly modified proofs only) for fuzzy relations too. Thus it is not necessary to write down these results for fuzzy relations again; nevertheless we will use them in the following. So we start by first defining some specific notions of relation theory.

Definition 2.8 *For each t-norm* **t** *and all fuzzy relations R, S let be the* domain, *the* range, *the* inverse relation *of R and the* (relational) product *of R, S:*

$$\begin{aligned} \mathrm{dom}\,(R) &=_{def} \{x \,\|\, \exists y((x,y)\,\varepsilon\, R)\}, \\ \mathrm{rg}\,(R) &=_{def} \{y \,\|\, \exists x((x,y)\,\varepsilon\, R)\}, \\ R^{-1} &=_{def} \{(x,y) \,\|\, (y,x)\,\varepsilon\, R\}, \\ R \circ_{\mathbf{t}} S &=_{def} \{(x,y) \,\|\, \exists z((x,z)\,\varepsilon\, R \wedge_{\mathbf{t}} (z,y)\,\varepsilon\, S)\}. \end{aligned}$$

Proposition 2.12 *For all fuzzy relations R, S one has*

$$\begin{aligned} &(i) && \mathrm{dom}\,(R^{-1}) = \mathrm{rg}\,(R) \quad \textit{and} \quad \mathrm{rg}\,(R^{-1}) = \mathrm{dom}\,(R), \\ &(ii) && \models \mathrm{dom}\,(R \circ_{\mathbf{t}} S) \subseteqq \mathrm{dom}\,(R), \\ &(iii) && \models \mathrm{rg}\,(R \circ_{\mathbf{t}} S) \subseteqq \mathrm{rg}\,(R), \\ &(iv) && (R^{-1})^{-1} = R, \\ &(v) && (R \circ_{\mathbf{t}} S)^{-1} = S^{-1} \circ_{\mathbf{t}} R^{-1}. \end{aligned}$$

Proof. All the results can be proved by routine methods; thus e.g. (v) essentially follows from the definition of the inverse fuzzy relation R^{-1} and the commutativity of $\wedge_t$ by

$$\begin{aligned}(R \circ_t S)^{-1} &= \{(x,y) \,\|\, \exists z((y,z)\,\varepsilon\, R \wedge_t (z,x)\,\varepsilon\, S)\} \\ &= \{(x,y) \,\|\, \exists z((x,z)\,\varepsilon\, S^{-1} \wedge_t (z,y)\,\varepsilon\, R^{-1})\} \\ &= S^{-1} \circ_t R^{-1}.\end{aligned}$$

And for (ii), for example, we have to show

$$\models \forall x(x \,\varepsilon\, \mathrm{dom}\,(R \circ_t S) \rightarrow x \,\varepsilon\, \mathrm{dom}\,(R))$$

which goes to show that

$$\models \forall x(\exists y((x,y)\,\varepsilon\, R \circ_t S) \rightarrow \exists z((x,z)\,\varepsilon\, R)), \tag{2.25}$$

but because of

$$[\![(x,y)\,\varepsilon\, R \circ_t S]\!] = [\![\exists z((x,z)\,\varepsilon\, R \wedge_t (z,y)\,\varepsilon\, S)]\!]$$

we have by monotonicity of the t-norms

$$\models \exists y((x,y)\,\varepsilon\, R \circ_t S) \rightarrow \exists y \exists z((x,z)\,\varepsilon\, R)$$

and hence (2.25) by first dropping the empty quantification $\exists y$ in the succedent and then universally quantifying. QED

Proposition 2.13 *If* LSC(t), *or if* $\mathcal{X}$ *is finite, or also if* R, S, T *have only a finite number of membership degrees, then it holds true that*

$$R \circ_t (S \circ_t T) = (R \circ_t S) \circ_t T.$$

Proof. The way of proving this equation is routine. The essential point is to prove that the formulas

$$\exists u((x,u)\,\varepsilon\, R \wedge_t \exists v((u,v)\,\varepsilon\, S \wedge_t (v,y)\,\varepsilon\, T))$$

and

$$\exists v(\exists u((x,u)\,\varepsilon\, R \wedge_t (u,v)\,\varepsilon\, S) \wedge_t (v,y)\,\varepsilon\, T)$$

have the same truth degrees. To get that result we have – besides using the associativity of $\wedge_t$ – first to "move the quantifier $\exists v$ outside" a conjunction and then to "move $\exists u$ inside" a corresponding conjunction. As stated in proposition 1.28 (i) this is possible without a change of truth degrees in the case of $\mathsf{LSC}(\boldsymbol{t})$. Obviously, these changes of quantifiers are unproblematic in those cases where the existential quantifiers $\exists u, \exists v$ correspond to the operations of taking a maximum of truth degrees, which happens if only a finite number of truth degrees are possible for the formulas following $\exists u$ or $\exists v$, and this happens e.g. for finite universes of discourse $\mathcal{X}$ and for those fuzzy relations R, S, T with only a finite number of membership degrees really appearing, i.e. for those membership functions μ_R, μ_S, μ_T of fuzzy relations which have, as usual functions, finite ranges. QED

Unfortunately, in general we cannot avoid the assumption $\mathsf{LSC}(\boldsymbol{t})$ in this proposition. To get a counterexample in the case that $\mathsf{LSC}(\boldsymbol{t})$ fails, let us choose $\boldsymbol{t} = \boldsymbol{t}_D =$ "drastic product" and as universe of discourse $\mathcal{X} = \mathbb{N} =$ "set of natural numbers (without zero!)". Consider the fuzzy relations

$$R = \langle\!\langle (1,5) \rangle\!\rangle_{0.9} \qquad S = \bigcup_{n=1}^{\infty} \langle\!\langle (5,n) \rangle\!\rangle_{(1-\frac{1}{n})}, \qquad T = \bigcup_{n=1}^{\infty} \langle\!\langle (n,1) \rangle\!\rangle_{1}.$$

Writing $R \circ_D S$ instead of $R \circ_{\boldsymbol{t}_D} S$ we get successively

$$S \circ_D T = \langle\!\langle (5,1) \rangle\!\rangle_1, \qquad R \circ_D (S \circ_D T) = \langle\!\langle (1,1) \rangle\!\rangle_{0.9}$$

and on the other side

$$R \circ_D S = \emptyset, \qquad (R \circ_D S) \circ_D T = \emptyset.$$[6]

As a further remark let us add here that the assumptions on the finiteness of $\mathcal{X}$ or of the set of membership degrees of R, S, T may also in the following propositions be added instead of $\mathsf{LSC}(\boldsymbol{t})$. However, for simplicity we mention only the case of $\mathsf{LSC}(\boldsymbol{t})$.

Proposition 2.14 *Consider t-norms $\boldsymbol{t}$ and $\boldsymbol{t}_1$ such that $\boldsymbol{t}$ distributes over $\boldsymbol{t}_1$, and a t-conorm $\boldsymbol{s}_{\boldsymbol{t}_2}$ such that $\boldsymbol{t}$ also distributes over $\boldsymbol{s}_{\boldsymbol{t}_2}$. It then holds*

[6] By the way, if we consider additionally the t-conorm $\vee = \max$, then we have for all $u, v, w \in [0,1]$

$$(u \vee v)\, \boldsymbol{t}_D\, w = (u\, \boldsymbol{t}_D\, w) \vee (v\, \boldsymbol{t}_D\, w)$$

and hence property (1) of OVCHINNIKOV (1981). Therefore our example too is a counterexample to the unrestricted associativity of the composition of fuzzy relations (as stated erroneously in OVCHINNIKOV (1981; p. 172)).

true that

$$(i) \quad \models R \circ_{\mathbf{t}} (S \cap_{\mathbf{t}_1} T) \subsetneqq (R \circ_{\mathbf{t}} S) \cap_{\mathbf{t}_1} (R \circ_{\mathbf{t}} T),$$
$$(ii) \quad \models R \circ_{\mathbf{t}} (S \cup_{\mathbf{t}_2} T) \subsetneqq (R \circ_{\mathbf{t}} S) \cup_{\mathbf{t}_2} (R \circ_{\mathbf{t}} T).$$

Proof. By distributivity of $\mathbf{t}$ over $\mathbf{t}_1$ we get

$$\begin{aligned} & R \circ_{\mathbf{t}} (S \cap_{\mathbf{t}_1} T) \\ = \; & \{(x,y) \,\|\, \exists z((x,z) \,\varepsilon\, R \wedge_{\mathbf{t}} ((z,y) \,\varepsilon\, S \wedge_{\mathbf{t}_1} (z,y) \,\varepsilon\, T))\} \\ = \; & \{(x,y) \,\|\, \exists z(((x,z) \,\varepsilon\, R \wedge_{\mathbf{t}} (z,y) \,\varepsilon\, S) \wedge_{\mathbf{t}_1} ((x,z) \,\varepsilon\, R \wedge_{\mathbf{t}} (z,y) \,\varepsilon\, T))\} \end{aligned}$$

which gives (i) by the definition of the union of fuzzy relations and the fact that

$$[\![\exists z(H_1 \wedge_{\mathbf{t}_1} H_2)]\!] \leq [\![\exists z H_1 \wedge_{\mathbf{t}_1} \exists z H_2]\!]$$

for all formulas H_1, H_2 and t-norms $\mathbf{t}_1$.

(ii) is proved by corresponding calculations now using the fact that

$$[\![\exists z(H_1 \vee_{\mathbf{t}_2} H_2)]\!] \leq [\![\exists z H_1 \vee_{\mathbf{t}_2} \exists z H_2]\!]$$

for all formulas H_1, H_2 and t-conorms $\mathbf{s}_{\mathbf{t}_2}$. QED

If we take $\mathbf{t} = \mathbf{t}_1$ in this proposition, the condition of the distributivity of $\mathbf{t}$ over $\mathbf{t}$ equivalently reduces to the idempotency of $\mathbf{t}$. For, clearly, if $\mathbf{t}$ is idempotent, then $\mathbf{t}$ distributes over itself, i.e.

$$u\,\mathbf{t}\,(v\,\mathbf{t}\,w) = (u\,\mathbf{t}\,v)\,\mathbf{t}\,(u\,\mathbf{t}\,w) \tag{2.26}$$

always holds true; on the other hand if (2.26) always holds true then put $v = w = 1$ which gives $u = u\,\mathbf{t}\,u$ and hence the idempotency of $\mathbf{t}$ which finally means $\mathbf{t} = \min$.

Proposition 2.15 *For any t-norm $\mathbf{t}$ with property* $\mathsf{LSC}(\mathbf{t})$ *the generalized monotonicity of the relational product of fuzzy relations holds true in the forms*

$$(i) \quad \models R \subsetneqq_{\mathbf{t}} S \rightarrow R \circ_{\mathbf{t}} T \subsetneqq_{\mathbf{t}} S \circ_{\mathbf{t}} T,$$
$$(ii) \quad \models S \subsetneqq_{\mathbf{t}} T \rightarrow R \circ_{\mathbf{t}} S \subsetneqq_{\mathbf{t}} R \circ_{\mathbf{t}} T,$$
$$(iii) \quad \models R_1 \subsetneqq_{\mathbf{t}} R_2 \wedge_{\mathbf{t}} S_1 \subsetneqq_{\mathbf{t}} S_2 \rightarrow_{\mathbf{t}} R_1 \circ_{\mathbf{t}} S_1 \subsetneqq_{\mathbf{t}} R_2 \circ_{\mathbf{t}} S_2.$$

Proof. We consider only (i), case (ii) is analogous. Suppose $\mathsf{LSC}(t)$. Then we have

$$\models R \subseteqq_t S \rightarrow \forall x \forall y \forall z((x,y)\,\varepsilon\, R \wedge_t (y,z)\,\varepsilon\, T \rightarrow_t (x,y)\,\varepsilon\, S \wedge_t (y,z)\,\varepsilon\, T)$$

because of proposition 1.18 (i) one always has

$$\models (H_1 \rightarrow_t H_2) \rightarrow (H_1 \wedge_t G \rightarrow_t H_2 \wedge_t G),$$

furthermore obviously

$$\models R \subseteqq_t S \rightarrow \forall x \forall y \forall z((x,y)\,\varepsilon\, R \wedge_t (y,z)\,\varepsilon\, T \rightarrow_t (x,z)\,\varepsilon\, S \circ_t T)$$

and therefore, using the fact that in the case of $\mathsf{LSC}(t)$

$$\models \forall x \forall y \forall z((x,y)\,\varepsilon\, R \wedge_t (y,z)\,\varepsilon\, T \rightarrow_t (x,z)\,\varepsilon\, S \circ_t T) \rightarrow \forall x \forall z(\exists y((x,y)\,\varepsilon\, R \wedge_t (y,z)\,\varepsilon\, T) \rightarrow_t (x,z)\,\varepsilon\, S \circ_t T)$$

(cf. proposition 1.27 (ii)), we get by transitivity of $\rightarrow$

$$\models R \subseteqq_t S \rightarrow \forall x \forall z((x,y)\,\varepsilon\, R \circ_t T \rightarrow_t (x,z)\,\varepsilon\, S \circ_t T)$$

which had to be proved.

To get (iii) we have to use (i), (ii) together with the transitivity of fuzzy inclusion for fuzzy relations:

$$\models R \subseteqq_t S \wedge_t S \subseteqq_t T \rightarrow_t R \subseteqq_t T$$

which is nothing else than a special case of the corresponding result for fuzzy sets in general. QED

2.3 The full image under a relation

In classical set theory, having given a set of ordered pairs, one always has two ways of looking at it: one may consider this set as a (binary) relation or as a (unary) mapping.[7] Depending on that point of view one is interested in different properties of and operations with sets of ordered pairs.

---

[7] We follow the usage to speak of a mapping also in the case that there is no uniqueness with respect to the second argument, mappings are thus "multiple valued" functions. We reserve the name of a "function" as usual for the unique mappings.

The same situation, of course, appears with fuzzy sets of ordered pairs. So far we have looked at them as fuzzy relations. For the purpose of the present book this perspective remains the essential one. Nevertheless, there is one notion from the realm of the mapping-related ones that has to be considered together with fuzzy relations: the full image of a (fuzzy) set under a (fuzzy) relation. And the background for this interest is in the later use of the so-called "compositional rule of inference" in connection with our discussions of fuzzy controllers.

From a more set theoretical point of view the notions and facts from the mapping-related world are as essential and interesting as those from the relation-related world. This also holds true in the field of fuzzy sets. Moreover, in the mapping-related world of fuzzy sets there arise completely new problems, e.g. at first giving a satisfactory definition of uniqueness that is nontrivial in the sense that it refers to an equality predicate that allows for truth degrees different from the degrees 1 and 0.[8]

Definition 2.9 *For each t-norm* $\mathbf{t}$*, every fuzzy relation* $R \in \mathbb{F}(\mathcal{X} \times \mathcal{Y})$ *and all fuzzy sets* $A \in \mathbb{F}(\mathcal{X})$ *and* $B \in \mathbb{F}(\mathcal{Y})$*, let*

$$R''A =_{def} \{y \,\|\, \exists x (x \,\varepsilon\, A \wedge_{\mathbf{t}} (x,y) \,\varepsilon\, R)\},$$

and in the case that $\mathbf{t}$ *has the property* $\mathsf{LSC}(\mathbf{t})$ *also*

$$R \downarrow B =_{def} \{x \,\|\, \forall y ((x,y) \,\varepsilon\, R \rightarrow_{\mathbf{t}} y \,\varepsilon\, B)\}.$$

Obviously, $R''A$ is a fuzzified version of the full image of a set under a mapping, i.e. under a binary relation. Simultaneously, $R''A$ corresponds to what BELLMAN/ZADEH (1970) called "conditioned fuzzy set". And $R \downarrow A$ corresponds to the @-operation of SANCHEZ (1984). In some (weak) sense this operation $R \downarrow B$ is dual to taking the full image $R''A$ of A under R.

As a side remark let us mention that this fuzzified full image of a fuzzy set under a fuzzy relation could be used to define the range and hence, cf. proposition 2.12 (i), also the domain of a fuzzy relation. The reason is that

$$rg(R) = R''(X^{[1]}) \tag{2.27}$$

holds true for any fuzzy relation $R \in \mathbb{F}(\mathcal{X} \times \mathcal{Y})$.

Usually the t-norm used in the definition of $R''A$ is clear from the context and the same one which is (essentially) used there. Thus we shall not mention it explicitely in our notation.

[8] For work in this direction the interested reader is referred to GOTTWALD (1980) as one possible approach.

Proposition 2.16 *For all fuzzy sets $A, B \in \mathbb{F}(\mathcal{X})$ and each fuzzy relation $R \in \mathbb{F}(\mathcal{X} \times \mathcal{Y})$ there hold true*

$$(i) \qquad \models R''(A \cup B) \equiv R''A \cup R''B,$$
$$(ii) \qquad \models R''(A \cap B) \subseteqq R''A \cap R''B;$$

in the case that $R''..$ is defined using the t-norm $t_G = \min$ it also holds true

$$(iii) \qquad \models R''A \cup_t R''B \subseteqq (R \cup_t R)''(A \cup_t B).$$

Proof. (i) By definition 2.9 and the distributivity of any t-norm t over $\vee$ according to proposition 1.12 one has

$$\begin{aligned} R''A \cup R''B &= \{y \| \exists x(x \varepsilon A \wedge_t (x,y) \varepsilon R) \vee \exists x(x \varepsilon B \wedge_t (x,y) \varepsilon R)\} \\ &= \{y \| \exists x((x \varepsilon A \wedge_t (x,y) \varepsilon R) \vee (x \varepsilon B \wedge_t (x,y) \varepsilon R))\} \\ &= \{y \| \exists x((x \varepsilon A \vee x \varepsilon B) \wedge_t (x,y) \varepsilon R)\} \\ &= R''(A \cup B). \end{aligned}$$

(ii) follows by the same type of calculations but now using proposition 1.26 (ii) for the step from the first to the second line, giving the (nonfuzzy) inclusion "$\supset$" instead of equality.

(iii) With reference to proposition 1.26 (iv) the calculations to determine $R''A \cup_t R''B$ can start as in case (i) and proceed as in (ii), but now with an inclusion "$\subset$" resulting, and have later on to use the distributivity of any t-conorm over $\wedge$ and that therefore one has

$$\models (H_1 \wedge H_3) \vee_t (H_2 \wedge H_3) \leftrightarrow (H_1 \vee_t H_2) \wedge (H_3 \vee_t H_3)$$

for all expressions H_1, H_2, H_3. The rest is routine matter. QED

The inspection of the proofs for the claims (i) and (ii) of this proposition indicates a further possibility of generalization: under some suitable assumptions $\cup$ and $\cap$ can be replaced by some $\cup_t$ or some $\cap_t$ respectively. We state the result without further proof.

Proposition 2.17 *Consider fuzzy sets $A, B \in \mathbb{F}(\mathcal{X})$ and a fuzzy relation $R \in \mathbb{F}(\mathcal{X} \times \mathcal{Y})$. Suppose that the definition of $R''..$ uses a t-norm t_1 and that another t-norm t is considered too.*

(i) If t_1 distributes over s_t then it holds true

$$\models R''(A \cup_t B) \subseteqq R''A \cup_t R''B.$$

(ii) If $\mathbf{t}_1$ distributes over $\mathbf{t}$ or if at least the subdistributivity condition $u\,\mathbf{t}_1\,(v\,\mathbf{t}\,w) \leq (u\,\mathbf{t}_1\,v)\,\mathbf{t}\,(u\,\mathbf{t}_1\,w)$ holds true for all $u, v, w \in [0,1]$, then one has

$$\models R''(A \cap_{\mathbf{t}} B) \subseteqq R''A \cap_{\mathbf{t}} R''B.$$

For the following considerations let us suppose for simplicity that $\mathcal{X} = \mathcal{Y}$ and thus all fuzzy sets are from $\mathbb{F}(\mathcal{X})$ and all fuzzy relations from $\mathbb{F}(\mathcal{X} \times \mathcal{X})$.

Proposition 2.18 *Suppose* $\mathsf{LSC}(\mathbf{t})$. *Then it holds true for all fuzzy sets A, B and all fuzzy relations R, S that*

$$
\begin{array}{ll}
(i) & \models A \subseteqq_{\mathbf{t}} B \rightarrow R''A \subseteqq_{\mathbf{t}} R''B, \\
(ii) & \models R \subseteqq_{\mathbf{t}} S \rightarrow R''A \subseteqq_{\mathbf{t}} S''A, \\
(iii) & \models A \subseteqq_{\mathbf{t}} B \rightarrow R \downarrow A \subseteqq_{\mathbf{t}} R \downarrow B, \\
(iv) & \models R \subseteqq_{\mathbf{t}} S \rightarrow S \downarrow A \subseteqq_{\mathbf{t}} R \downarrow A.
\end{array}
$$

Proof. (i) Because of proposition 1.13 the logical validity of a generalized implication is equivalent to an inequality of the corresponding truth degrees. What has to be shown therefore is that

$$[\![A \subseteqq_{\mathbf{t}} B]\!] \leq [\![R''A \subseteqq_{\mathbf{t}} R''B]\!].$$

We start from the right hand side of this inequality and have by the corresponding definitions

$$
\begin{aligned}
&[\![R''A \subseteqq_{\mathbf{t}} R''B]\!] \\
&\quad = [\![\forall x(x \,\varepsilon\, R''A \rightarrow_{\mathbf{t}} x \,\varepsilon\, R''B)]\!] \\
&\quad = [\![\forall x(\exists y(y \,\varepsilon\, A \wedge_{\mathbf{t}} (y,x) \,\varepsilon\, R) \rightarrow_{\mathbf{t}} \exists z(z \,\varepsilon\, B \wedge_{\mathbf{t}} (z,x) \,\varepsilon\, R))]\!]
\end{aligned}
$$

and using proposition 1.27 (ii) furthermore

$$
\begin{aligned}
&[\![R''A \subseteqq_{\mathbf{t}} R''B]\!] \\
&\quad = [\![\forall x \forall y((y \,\varepsilon\, A \wedge_{\mathbf{t}} (y,x) \,\varepsilon\, R) \rightarrow_{\mathbf{t}} \exists z(z \,\varepsilon\, B \wedge_{\mathbf{t}} (z,x) \,\varepsilon\, R))]\!].
\end{aligned}
$$

Now from (1.2) together with (Φ1) and proposition 1.18 (i) we successively get

$$
\begin{aligned}
&[\![R''A \subseteqq_{\mathbf{t}} R''B]\!] \\
&\quad \geq [\![\forall x \forall y((y \,\varepsilon\, A \wedge_{\mathbf{t}} (y,x) \,\varepsilon\, R) \rightarrow_{\mathbf{t}} (y \,\varepsilon\, B \wedge_{\mathbf{t}} (y,x) \,\varepsilon\, R))]\!] \\
&\quad \geq [\![\forall x \forall y(y \,\varepsilon\, A \rightarrow_{\mathbf{t}} y \,\varepsilon\, B)]\!]
\end{aligned}
$$

which by dropping the empty quantification "$\forall x$" just means

$$[\![R''A \subseteqq_t R''B]\!] \geq [\![A \subseteqq_t B]\!].$$

The same kind of calculations yield (ii). We omit these details.

(iii) As before it has to be shown that the inequality

$$[\![A \subseteqq_t B]\!] \leq [\![R \downarrow A \subseteqq_t R \downarrow B]\!]$$

holds true. But now the corresponding definitions lead to the following estimations

$$\begin{aligned}
&[\![R \downarrow A \subseteqq_t R \downarrow B]\!] \\
&\quad = [\![\forall x(\forall y((x,y)\,\varepsilon\, R \rightarrow_t y\,\varepsilon\, A) \rightarrow_t \forall z((x,z)\,\varepsilon\, R \rightarrow_t z\,\varepsilon\, B))]\!] \\
&\quad = [\![\forall x \forall z(\forall y((x,y)\,\varepsilon\, R \rightarrow_t y\,\varepsilon\, A) \rightarrow_t ((x,z)\,\varepsilon\, R \rightarrow_t z\,\varepsilon\, B))]\!] \\
&\quad \geq [\![\forall x \forall z(((x,z)\,\varepsilon\, R \rightarrow_t z\,\varepsilon\, A) \rightarrow_t ((x,z)\,\varepsilon\, R \rightarrow_t z\,\varepsilon\, B))]\!] \\
&\quad \geq [\![\forall x \forall z(z\,\varepsilon\, A) \rightarrow_t z\,\varepsilon\, B)]\!]
\end{aligned}$$

which use the antimonotonicity of φ_t in the first argument, i.e. proposition 1.4 (i), as well as proposition 1.18 (ii).

Finally (iv) can be proven along the same line of arguments. But now instead of proposition 1.18 (ii) reference has to be made to proposition 1.18 (iii) which causes the reversal in the order of appearance of R and S on both sides of the arrow $\rightarrow$ in the formula. QED

2.4 Special types of fuzzy relations

In order to be able to discuss some special types of fuzzy relations in the following, we first have to introduce some specific properties of fuzzy relations. In this section these shall always be crisp properties, i.e. properties which a fuzzy relation either has or does not have. Different lists of such properties can be found in for example GOTTWALD (1974), DUBOIS/PRADE (1980), OVCHINNIKOV (1981), and CHAKRABORTY/DAS (1983). As we only consider binary relations, we will change our notation slightly and write – as for usual binary relations – simply xRy for the truth degree $(x,y)\,\varepsilon\, R$, i.e. we use $[\![xRy]\!] = \mu_R(x,y)$.

Definition 2.10 *A fuzzy relation R is (with respect to a given t-norm t and a negation function n)*

(i)	*reflexive (in $\mathcal{X}$)*	*iff*	$\models \forall x(xRx)$,
(ii)	*irreflexive*	*iff*	$\models \forall x \sim_n (xRx)$,
(iii)	*t-transitive*	*iff*	$\models \forall x \forall y \forall z(xRy \wedge_t yRz \rightarrow_t xRz)$,
(iv)	*symmetric*	*iff*	$\models \forall x \forall y(xRy \rightarrow_t yRx)$,
(v)	*t-antisymmetric*	*iff*	$\models \forall x \forall y(xRy \wedge_t yRx \rightarrow_t x \doteq y)$,
(vi)	*t-asymmetric*	*iff*	$\models \forall x \forall y \sim_n (xRy \wedge_t yRx)$.

Remark. By the usual truth condition (1.6) for logical validity and definition (1.1) of the many-valued universal quantifier obviously the irreflexivity condition becomes: $[\![xRx]\!] = 1$ for each $x \in \mathcal{X}$. Therefore the definition 2.10 (ii) of irreflexivity and the definition 2.10 (vi) of t-asymmetry are *independent* of the choice of a specific negation function.

As an additional consequence the symmetry condition (iv) becomes that: $[\![xRy \rightarrow_t yRx]\!] = 1$ for all $x, y \in \mathcal{X}$. As long as one supposes that the t-norm t which is involved here has the property $\mathsf{LSC}(t)$, the condition reads: $[\![xRy]\!] \leq [\![yRx]\!] = 1$ for all $x, y \in \mathcal{X}$. Thus the definition 2.10 (iv) is also *independent* of the choice of a specific lower semicontinuous t-norm.

As usual these properties may also be characterized using the set theoretic operations for fuzzy relations. To do this we need a kind of fuzzified diagonal. Let the *fuzzy diagonal* of $\mathcal{X}$ be the fuzzy relation

$$\Delta_{\mathcal{X}} =_{def} \bigcup_{a \in \mathcal{X}} \langle\!\langle (a,a) \rangle\!\rangle_1 = \{(x,y) \in \mathcal{X} \times \mathcal{X} \,\|\, x \doteq y\}. \tag{2.28}$$

Proposition 2.19 *For each fuzzy relation R it holds true that*

(i)	*R reflexive (in $\mathcal{X}$)*	*iff*	$\models \Delta_{\mathcal{X}} \subseteqq R$,
(ii)	*R irreflexive*	*iff*	$\models R \cap_t \Delta_{\mathcal{X}} \equiv \emptyset$,
(iii)	*R t-transitive*	*iff*	$\models R \circ_t R \subseteqq_t R$
		iff	$\models R \circ_t R \subseteqq R$,
(iv)	*R symmetric*	*iff*	$\models R \subseteqq R^{-1}$,
(v)	*R t-antisymmetric*	*iff*	$\models R \cap_t R^{-1} \subseteqq \Delta_{\mathcal{X}}$,
(vi)	*R t-asymmetric*	*iff*	$\models R \cap_t R^{-1} \equiv \emptyset$.

Proof. Obvious.

First we look for a fuzzification of equivalence relations. Such fuzzified equivalence relations have been discussed, for example by KLAUA (1970),

ZADEH (1971), OVCHINNIKOV (1981) and CHAKRABORTY/DAS (1983a). We will give only the most basic facts.

Definition 2.11 *A* fuzzy equivalence relation R *(in $\mathcal{X}$) is such a fuzzy relation which is reflexive (in $\mathcal{X}$), symmetric and* **t**-*transitive. For each fuzzy equivalence relation R and each $b \in \mathcal{X}$ the* fuzzy R-equivalence class *of b is*

$$\langle b \rangle_R =_{def} \{x \parallel xRb\}.$$

By the symmetry of each fuzzy equivalence relation it is immediately

$$\langle b \rangle_R = \{x \parallel bRx\}$$

and using the notion of the full image from definition 2.9 we hence have

$$\langle b \rangle_R = R'' \langle\!\langle b \rangle\!\rangle_1$$

for the fuzzy R-equivalence classes.

Proposition 2.20 *For each fuzzy equivalence relation R, each t-norm* **t** *with the property* LSC(**t**) *and all $a, b \in \mathcal{X}$, there hold true:*

$$\begin{array}{ll} (i) & \models a \,\varepsilon\, \langle a \rangle_R, \\ (ii) & \models b \,\varepsilon\, \langle a \rangle_R \leftrightarrow_{\mathbf{t}} aRb, \\ (iii) & \models \langle a \rangle_R \equiv_{\mathbf{t}} \langle b \rangle_R \leftrightarrow_{\mathbf{t}} aRb, \\ (iv) & \models \exists x (x \,\varepsilon\, \langle a \rangle_R \cap_{\mathbf{t}} \langle b \rangle_R) \leftrightarrow_{\mathbf{t}} aRb. \end{array}$$

Proof. (ii) is simply a restatement of the definition of $\langle a \rangle_R$; and (i) follows from this definition by reflexivity of R.

(iii) From definition of $\equiv_{\mathbf{t}}$ and from (ii) one immediately has

$$\models \langle a \rangle_R \equiv_{\mathbf{t}} \langle b \rangle_R \rightarrow_{\mathbf{t}} (aRa \leftrightarrow_{\mathbf{t}} aRb)$$

and hence by reflexivity of R

$$\models \langle a \rangle_R \equiv_{\mathbf{t}} \langle b \rangle_R \rightarrow_{\mathbf{t}} aRb;$$

furthermore by transitivity of R and symmetry

$$\models aRb \rightarrow_{\mathbf{t}} (xRa \rightarrow_{\mathbf{t}} xRb),$$
$$\models aRb \rightarrow_{\mathbf{t}} (xRb \rightarrow_{\mathbf{t}} xRa).$$

Therefore proposition 1.6 (i) immediately gives

$$\models aRb \rightarrow_t ((xRa \rightarrow_t xRb) \wedge (xRb \rightarrow_t xRa)).$$

Because of the linear ordering of the set of truth degrees and of proposition 1.4 (ii) and condition (T3) for all formulas H, G one has

$$[\![(H \rightarrow_t G) \wedge (G \rightarrow_t H)]\!] = [\![(H \rightarrow_t G) \wedge_t (G \rightarrow_t H)]\!]$$

which in the present case gives

$$\models aRb \rightarrow_t ((xRa \leftrightarrow_t xRb)$$

and hence, e.g. using proposition 1.27 (i),

$$\models aRb \rightarrow_t \langle a \rangle_R \equiv_t \langle b \rangle_R.$$

Thus (iii) is proved.

(iv) By (i) and (ii) it holds true that

$$\models aRb \rightarrow_t b \,\varepsilon\, \langle a \rangle_R \cap_t \langle b \rangle_R$$

and hence of course as part of (iv) already

$$\models aRb \rightarrow_t \exists x (x \,\varepsilon\, \langle a \rangle_R \cap_t \langle b \rangle_R). \tag{2.29}$$

Otherwise we also have

$$\models x \,\varepsilon\, \langle a \rangle_R \cap_t \langle b \rangle_R \rightarrow_t xRa \wedge_t xRb$$

and thus together with t-transitivity and symmetry of R

$$\models x \,\varepsilon\, \langle a \rangle_R \cap_t \langle b \rangle_R \rightarrow_t aRb$$

which together with LSC(t) gives

$$\models \exists x (x \,\varepsilon\, \langle a \rangle_R \cap_t \langle b \rangle_R \rightarrow_t aRb). \tag{2.30}$$

Thus finally (iv) is proved by (2.29) and (2.30). QED

Our next concern will be the connection of fuzzy equivalence relations with generalized partitions of the universe of discourse.

Definition 2.12 *A* fuzzy partition *of (the universe of discourse) $\mathcal{X}$ is such a crisp (!) class $\mathcal{Z}$ of fuzzy subsets of $\mathcal{X}$ such that*

$$\text{(Z1)} \qquad \bigcup_{A \in \mathcal{Z}} A = X$$

and for all $A, B \in \mathcal{Z}$:

$$\text{(Z2)} \qquad \models \exists x (x \varepsilon A \cap_t B) \rightarrow_t A \equiv_t B.$$

Proposition 2.21 *Suppose* $\mathsf{LSC}(t)$.

(i) If R is a fuzzy equivalence relation (in the universe $\mathcal{X}$), then the set of R-equivalence classes

$$\mathcal{Z}/R = \{\langle a \rangle_R \mid a \in \mathcal{X}\}$$

is a fuzzy partition of $\mathcal{X}$.

(ii) If $\mathcal{Z}$ is a fuzzy partition of $\mathcal{X}$, then a fuzzy equivalence relation (in $\mathcal{X}$) is defined by

$$R_{\mathcal{Z}} = \{(a,b) \parallel \mathbb{W}_{A \in \mathcal{Z}} \mathbb{W}_{B \in \mathcal{Z}} (a \varepsilon A \wedge_t b \varepsilon B \wedge_t A \equiv_t B)\}.$$

Proof. (i) Because from proposition 2.20 (i) one has $[\![a \varepsilon \langle a \rangle_R]\!] = 1$ for each $a \in \mathcal{X}$, we have that R has property (Z1). Property (Z2) is a straightforward consequence of proposition 2.20 (iii) and (iv).

(ii) Obviously $R_{\mathcal{Z}}$ is symmetric. For the reflexivity of $R_{\mathcal{Z}}$ one has to consider

$$\begin{aligned} [\![(a,a) \varepsilon R_{\mathcal{Z}}]\!] &= [\![\mathbb{W}_{A,B \in \mathcal{Z}} (a \varepsilon A \wedge_t a \varepsilon B \wedge_t A \equiv_t B)]\!] \\ &= [\![\mathbb{W}_{A \in \mathcal{Z}} (a \varepsilon A \wedge_t a \varepsilon A)]\!] \end{aligned}$$

and hence to prove that

$$\sup_{A \in \mathcal{Z}} [\![a \varepsilon A \wedge_t a \varepsilon A]\!] = 1. \tag{2.31}$$

But for (2.31) it is necessary and sufficient to have

$$\sup_{A \in \mathcal{Z}} [\![a \varepsilon A]\!] = 1$$

and that is precisely condition (Z1).

In order to prove the t-transitivity of R_Z consider any $a, b, c \in \mathcal{X}$. Then

$$\begin{aligned}
&[\![(a,b)\,\varepsilon\, R_Z \wedge_t (b,c)\,\varepsilon\, R_Z]\!] = \\
&\quad [\![\textstyle\bigvee_{A,B,C,D\in Z}(a\,\varepsilon\, A \wedge_t b\,\varepsilon\, B \wedge_t A \equiv_t B \wedge_t \\
&\quad \wedge_t b\,\varepsilon\, D \wedge_t c\,\varepsilon\, C \wedge_t D \equiv_t C)]\!].
\end{aligned}$$

By (Z2) we easily get

$$[\![b\,\varepsilon\, B \wedge_t b\,\varepsilon\, D]\!] \leq [\![B \equiv_t D]\!]$$

and from this inequalitiy together with

$$[\![A \equiv_t B \wedge_t B \equiv_t D \wedge_t D \equiv_t C]\!] \leq [\![A \equiv_t C]\!]$$

by t-transitivity of $\equiv_t$ and (T2), i.e. monotonicity of the t-norm t,

$$\begin{aligned}
[\![(a,b)\,\varepsilon\, R_Z \wedge_t (b,c)\,\varepsilon\, R_Z]\!] &\leq [\![\textstyle\bigvee_{A,C\in Z}(a\,\varepsilon\, A \wedge_t c\,\varepsilon\, C \wedge_t A \equiv_t C)]\!] \\
&\leq [\![(a,c)\,\varepsilon\, R/Z]\!]
\end{aligned}$$

Hence the fuzzy relation R is transitive too. QED

Our approach to fuzzy equivalence relations – or: (fuzzy) similarity relations using the terminology of ZADEH (1971) – is a straightforward generalization of the corresponding approaches of ZADEH (1971) and of KLAUA (1970). The approach gives a fuzzification of the notion of equivalence relation in a given universe of discourse. What classically comes to the same is to consider an equivalence relation in some crisp set: one only has to interpret that crisp set as the universe of discourse.

Unfortunately, for fuzzy sets the situation is not so simple. It would be nice to have not only a theory of fuzzy equivalence relations in a given universe of discourse, but also a theory of fuzzy equivalence relations in a given *fuzzy* set A, in the sense that the fuzzy equivalence relation in A is connected with any kind of (generalized – i.e. fuzzy) partition of A. Yet, so far there has been no success in this problem.

As the next topic, let us discuss the construction of the transitive hull of a given fuzzy relation. As with fuzzy equivalence relations, the approach is mainly along the same lines as ZADEH (1971), but as before we get some more general results.

Definition 2.13 *For fuzzy relations R and any given t-norm $\mathbf{t}$ we put recursively for all natural numbers n*

$$R^1 =_{def} R,$$
$$R^{n+1} =_{def} R^n \circ_{\mathbf{t}} R,$$

and take as the **t**-transitive hull *of R the fuzzy relation*

$$\mathcal{T}_{\mathbf{t}}(R) =_{def} \bigcup_{n=1}^{\infty} R^n.$$

Proposition 2.22 *For each fuzzy relation R we have*

$$(i) \quad \models R \subseteq \mathcal{T}_{\mathbf{t}}(R),$$
$$(ii) \quad R \ \mathbf{t}\text{-transitive} \Rightarrow R = \mathcal{T}_{\mathbf{t}}(R),$$

and if the t-norm $\mathbf{t}$ has property $\mathsf{LSC}(\mathbf{t})$

$$(iii) \quad \mathcal{T}_{\mathbf{t}}(R) \ \text{is } \mathbf{t}\text{-transitive},$$
$$(iv) \quad \mathcal{T}_{\mathbf{t}}(\mathcal{T}_{\mathbf{t}}(R)) = \mathcal{T}_{\mathbf{t}}(R).$$

Proof. (i) is obvious. (ii) hence is proved if we show $\models \mathcal{T}_{\mathbf{t}}(R) \subseteq R$ for $\mathbf{t}$-transitive fuzzy relations R. But if R is $\mathbf{t}$-transitive, i.e. if $\models R \circ_{\mathbf{t}} R \subseteq R$, we have by $R^3 = R^2 \circ_{\mathbf{t}} R$ and by the transitivity of $\subseteq$ immediately from this $\mathbf{t}$-transitivity: $\models R^3 \subseteq R$. Recursively we get $\models R^n \subseteq R$ for each integer $n \geq 1$. Therefore obviously $\models \mathcal{T}_{\mathbf{t}}(R) \subseteq R$ and (ii) is proved.

For (iii) let m, n be any natural numbers. Using $\mathsf{LSC}(\mathbf{t})$ and the definition of R^m in a straightforward way, one gets

$$[\![(x,y) \varepsilon R^n \wedge_{\mathbf{t}} (y,z) \varepsilon R^m]\!] \leq [\![(x,z) \varepsilon R^{n+m}]\!]$$

and thus

$$\models (x,y) \varepsilon R^n \wedge_{\mathbf{t}} (y,z) \varepsilon R^m \rightarrow_{\mathbf{t}} (x,z) \varepsilon \mathcal{T}_{\mathbf{t}}(R).$$

Now taking suprema on all $n \geq 1$ and on all $m \geq 1$ gives by $\mathsf{LSC}(\mathbf{t})$

$$\models (x,y) \varepsilon \mathcal{T}_{\mathbf{t}}(R) \wedge_{\mathbf{t}} (y,z) \varepsilon \mathcal{T}_{\mathbf{t}}(R) \rightarrow_{\mathbf{t}} (x,z) \varepsilon \mathcal{T}_{\mathbf{t}}(R).$$

i.e. the $\mathbf{t}$-transitivity of $\mathcal{T}_{\mathbf{t}}(R)$. By (ii) and (iii) now (iv) is obvious. QED

We can prove some relatively weak results on the monotonicity of the $\mathcal{T}_{\mathbf{t}}$-operator for the general case of any t-norm. More interesting monotonicity properties are provable in the case that $\mathbf{t} = \min$.

Proposition 2.23 *For fuzzy relations R, S we have:*

(*i*) *if S is **t**-transitive and* $\models R \subseteqq S$, *then* $\models \mathcal{T}_{\mathbf{t}}(R) \subseteqq S$;

and with the additional assumption LSC($\mathbf{t}$)

(*ii*) *if* $\models R \subseteqq_{\mathbf{t}} S$, *then* $\models \mathcal{T}_{\mathbf{t}}(R) \subseteqq_{\mathbf{t}} \mathcal{T}_{\mathbf{t}}(S)$.

Proof. If S is $\mathbf{t}$-transitive then $\models S^n \subseteqq S$ for each $n \geq 1$ as mentioned in the foregoing proof. Furthermore

$$\models R^2 \subseteqq R \circ_{\mathbf{t}} S \quad \text{and} \quad \models R \circ_{\mathbf{t}} S \subseteqq S^2$$

by $\models R \subseteqq S$ and the monotonicity of $\circ_{\mathbf{t}}$. Hence together that gives

$$\models R^2 \subseteqq S;$$

and in the same way more generally

$$\models R^n \subseteqq S$$

for each $n \geq 1$. Therefore obviously

$$\models \mathcal{T}_{\mathbf{t}}(R) \subseteqq S$$

and hence (i).

By proposition 2.22 (i) we have $\models R \subseteqq \mathcal{T}_{\mathbf{t}}(S)$ from $\models R \subseteqq S$. Because of the $\mathbf{t}$-transitivity of the fuzzy relation $\mathcal{T}_{\mathbf{t}}(S)$ the result (ii) follows from part (i). QED

This proposition discusses the situation that, either with or without the $\mathbf{t}$-transitivity of the fuzzy relation S, one has that the graded inclusion "$R \subseteqq_{\mathbf{t}} S$" has truth degree 1. Then the statement is that some other inclusions have truth degree 1 too. More general and thus more interesting would be to have instead for example inequalities between the degrees to which these fuzzy inclusion relations hold true. Such results will only now be established for a special case.[9]

Proposition 2.24 *For the t-norm* $\mathbf{t} = \mathbf{t}_G = \min$ *one has for all fuzzy relations R, S in the case that S is $\mathbf{t}_G$-transitive:*

(*i*) $\models R \subseteqq_{\mathbf{t}_G} S \rightarrow \mathcal{T}_{\mathbf{t}_G}(R) \subseteqq_{\mathbf{t}_G} S$,

and without any transitivity restraint for S the generalized monotonicity

(*ii*) $\models R \subseteqq_{\mathbf{t}_G} S \rightarrow \mathcal{T}_{\mathbf{t}_G}(R) \subseteqq_{\mathbf{t}_G} \mathcal{T}_{\mathbf{t}_G}(S)$.

[9] At present it is an open problem whether the results of the next proposition can be extended to other t-norms besides t = min or not.

Proof. (i) Suppose S to be $\mathbf{t}_G$-transitive, i.e. to be min-transitive. Then one has (always deleting the index $\mathbf{t}_G$):

$$\models R \subseteqq S \to R^2 \subseteqq R \circ S$$
$$\models R \subseteqq S \to R \circ S \subseteqq S^2$$

and hence together with $\models S^2 \subseteqq S$:

$$\models R \subseteqq S \to R^2 \subseteqq S.$$

Continuing this line of reasoning gives

$$\models R \subseteqq S \to R^n \subseteqq S$$

for each $n \geq 1$. From this one gets together with

$$\inf_{n\geq 1} [\![R^n \subseteqq S]\!] = [\![\bigcup_{n=1}^{\infty} R^n \subseteqq S]\!]$$

the fact that

$$\models R \subseteqq S \to \mathcal{T}(R) \subseteqq S.$$

(ii) Again deleting all indices $\mathbf{t}_G$ we have from proposition 2.22 (i)

$$\models R \subseteqq S \to R \subseteqq S \wedge S \subseteqq \mathcal{T}(S)$$

and hence

$$\models R \subseteqq S \to R \subseteqq \mathcal{T}(S).$$

Now part (i) of our proposition gives because of the transitivity of $\mathcal{T}(S)$

$$\models R \subseteqq \mathcal{T}(S) \to \mathcal{T}(R) \subseteqq \mathcal{T}(S)$$

and thus

$$\models R \subseteqq S \to \mathcal{T}(R) \subseteqq \mathcal{T}(S)$$

altogether. QED

As a third topic let us discuss fuzzy ordering relations. Again following ZADEH (1971) and generalizing GOTTWALD (1974) we give the following

Definition 2.14 *A fuzzy relation which is reflexive and* **t***-transitive is called* fuzzy preordering; *and a reflexive,* **t***-transitive, and* **t***-antisymmetric fuzzy relation is called* fuzzy partial ordering.

For fuzzy relations and for crisp relations it is enough to discuss the reflexive case: in a simple way one can change from reflexive fuzzy ordering relations to corresponding irreflexive ones in essentially the same manner as in classical set theory.

Definition 2.15 *For each t-norm* $\mathbf{t}$ *and each fuzzy relation* $R \in \mathbb{F}(\mathcal{X} \times \mathcal{X})$ *let*

$$\begin{aligned} R^+ &=_{def} R \cup_{\mathbf{t}} \Delta_{\mathcal{X}}, \\ R^- &=_{def} R \cap_{\mathbf{t}} \complement \Delta_{\mathcal{X}}. \end{aligned}$$

Corollary 2.25 *For each t-norm* $\mathbf{t}$ *and each fuzzy relation* R *one has*

(*i*) R^- *is irreflexive,*

(*ii*) R^+ *is reflexive.*

Proof. Obvious.

Proposition 2.26 *Let* R *be any fuzzy relation and* $\mathbf{t}$ *any t-norm with the property* $\mathsf{LSC}(\mathbf{t})$. *Then one has*

(*i*) *If* R *is* **t**-*transitive then* R^+ *is* **t**-*transitive too.*

(*ii*) *If* R^- *is* **t**-*transitive then* R *is* **t**-*transitive too.*

(*iii*) *If* R *is* **t**-*transitive and* **t**-*antisymmetric then* R^- *is* **t**-*transitive.*

(*iv*) *If* R *is* **t**-*antisymmetric then* R^- *is* **t**-*asymmetric.*

Proof. Straightforward from the corresponding definitions.

For fuzzy ordering relations these results give the above-mentioned simple connection between the reflexive and the irreflexive case. To state that result we adopt here the

Definition 2.16 *A fuzzy relation* $R \in \mathbb{F}(\mathcal{X} \times \mathcal{X})$ *is an* irreflexive fuzzy partial ordering *iff* R *is an irreflexive,* **t**-*transitive and* **t**-*asymmetric relation.*

Proposition 2.27 *For each fuzzy relation* R *and any t-norm* $\mathbf{t}$ *with property* $\mathsf{LSC}(\mathbf{t})$ *one has:*

(i) *If R is a fuzzy partial ordering then R^- is an irreflexive fuzzy partial ordering.*

(ii) *If R is an irreflexive fuzzy partial ordering then R^+ is a fuzzy partial ordering.*

The proof is by straightforward calculations and will not be given in detail. But because of this result we confine ourselves to the case of reflexive fuzzy ordering relations.

To formulate some of our following results we refer to the terminology of for example ring theory and say that a t-norm $\boldsymbol{t}$ has *zero divisors* iff there exist $u, v \in (0,1]$ such that $u \,\boldsymbol{t}\, v = 0$; in such a case u, v itself are called zero divisors; cf. also OVCHINNIKOV (1981).

Obviously there are t-norms which have zero divisors. The Łukasiewicz t-norm $\boldsymbol{t}_L$ is an example: because of

$$\boldsymbol{t}_L(u, v) = \max\{0, u + v - 1\}$$

a pair (u_0, v_0) with $0 < u_0, v_0 \leq 1$ is a pair of zero divisors of $\boldsymbol{t}_L$ iff $u_0 + v_0 \leq 1$. In addition, the t-norm $\boldsymbol{t}_P$ is an example of a t-norm which does not have zero divisors.

Furthermore, the ordering relation (1.12) of t-norms gives additional information concerning t-norms with zero divisors. One easily recognizes that for all t-norms $\boldsymbol{t}_1, \boldsymbol{t}_2$

$$\boldsymbol{t}_1 \leqq \boldsymbol{t}_2 \quad \text{and} \quad \boldsymbol{t}_2 \text{ has zero divisors} \quad \Rightarrow \quad \boldsymbol{t}_1 \text{ has zero divisors} \tag{2.32}$$

holds true. Therefore also

$$\begin{aligned} \boldsymbol{t}_1 \leqq \boldsymbol{t}_2 \quad \text{and} \quad & \boldsymbol{t}_1 \text{ is without zero divisors} \\ \Rightarrow \quad & \boldsymbol{t}_2 \text{ is without zero divisors} \end{aligned} \tag{2.33}$$

All together that gives for each t-norm $\boldsymbol{t}$:

$$\boldsymbol{t} \leqq \boldsymbol{t}_L \quad \Rightarrow \quad \boldsymbol{t} \text{ has zero divisors,} \tag{2.34}$$

$$\boldsymbol{t}_P \leqq \boldsymbol{t} \quad \Rightarrow \quad \boldsymbol{t} \text{ is without zero divisors.} \tag{2.35}$$

Sometimes the considerations are restricted to Archimedean t-norms, cf. WEBER (1983). A t-norm $\boldsymbol{t}$ is called *Archimedean* iff $\boldsymbol{t}$ is a continuous function and for each $0 \neq u \in [0,1]$ one has $u \,\boldsymbol{t}\, u < u$. Each Archimedean t-norm $\boldsymbol{t}$ has a generating function $f : [0,1] \to [0,\infty]$ which is decreasing and continuous with $f(1) = 0$ such that always

$$u \,\boldsymbol{t}\, v = f^{(-1)}(f(u) + f(v)).$$

Here $f^{(-1)} : [0,\infty] \to [0,1]$ is the pseudo-inverse of f defined as

$$f^{(-1)}(y) =_{def} \begin{cases} f^{-1}(y), & \text{if } y \in rg(f), \\ 0 & \text{otherwise.} \end{cases}$$

Thus one has

$$u\, \mathbf{t}\, v = 0 \;\Leftrightarrow\; f(u) + f(v) \geq f(0).$$

That means that for each Archimedean t-norm $\mathbf{t}$ with generating function f one has

$$\mathbf{t} \text{ has zero divisors } \Leftrightarrow$$
$$f(0) \leq f(u_0) + f(v_0) \quad \text{for some } u_0, v_0 \in (0,1].$$

Occasionally the subclass of strict t-norms is also discussed. A t-norm is called *strict* iff it is Archimedean and strictly monotonously increasing in $[0,1]$. Then obviously one has for each t-norm $\mathbf{t}$:

$$\mathbf{t} \text{ strict } \Rightarrow \mathbf{t} \text{ without zero divisors.}$$

Proposition 2.28 *If R is a fuzzy partial ordering (preordering) then the crisp relation*

$$R^{\geq 1} = \{(x,y) \in \mathrm{supp}(R) \mid [\![xRy]\!] = 1\}$$

is a partial ordering (preordering) in the usual sense.

Proof. Straightforward from definitions.

Unfortunately a corresponding result for supp (R) instead of the kernel $R^{\geq 1}$ holds true only with restrictions on the t-norm used in the formulation of transitivity and antisymmetry properties.

Proposition 2.29 *If the t-norm $\mathbf{t}$ has zero divisors and $\mathcal{X}$ at least three elements, then there exists a fuzzy partial ordering R_0 such that the usual crisp relation* supp(R_0) *is not transitive.*

Proof. Consider $a,b,c \in \mathcal{X}$ and take zero divisors $u,v \in (0,1]$ such that $u\mathbf{t}v = 0$. Then

$$R_0 = \langle\!\langle (a,b) \rangle\!\rangle_u \cup \langle\!\langle (b,c) \rangle\!\rangle_v \cup \Delta_{\mathcal{X}}$$

is a fuzzy partial ordering, but the crisp relation supp $(R_0) = \{(a,b),(b,c)\}$ is not transitive (in the usual sense). QED

Additionally we mention that already in the case that card$(\mathcal{X}) \geq 2$ and t has zero divisors, a fuzzy partial ordering may be found whose support is not an antisymmetric relation (in the usual sense).

Proposition 2.30 *Suppose that the t-norm* t *does not have zero divisors. Then for each fuzzy partial ordering (preordering)* R *the support* supp(R) *is a crisp partial ordering (preordering) in the usual sense.*

Proof. Straightforward from definitions.

Proposition 2.31 (i) *If* R, S *are fuzzy partial orderings (preorderings) then for each t-norm* t *the fuzzy relation* $R \cap_t S$ *too is a fuzzy partial ordering (preordering).*

(ii) *Assume* $\mathsf{LSC}(t)$. *Consider any sequence* $(R_i)_{i<\alpha}$ *of type* α, α *any ordinal number, of fuzzy partial orderings (preorderings) such that for all indices* $i, j < \alpha$:

$$\text{if } i < j \text{ then } \models R_i \subseteq_t R_j.$$

Then the fuzzy relation

$$\bar{R} = \bigcup_{i=1}^{\alpha} R_i$$

is a fuzzy partial ordering (preordering) too.

Proof. (i) The reflexivity of $R \cap_t S$ is obvious. For the t-transitivity we have

$$\begin{aligned} &[\![x(R \cap_t S)y \wedge_t y(R \cap_t S)z]\!] \\ &\quad = [\![xRy \wedge_t xSy \wedge_t yRz \wedge_t ySz]\!] \\ &\quad \leq [\![xRz \wedge_t xSz]\!] = [\![x(R \cap_t S)z]\!] \end{aligned}$$

by the t-transitivity of R and S, and thus the t-transitivity of $R \cap_t S$ results. If both R, S are t-antisymmetric the t-antisymmetry of $R \cap_t S$ follows in the same way.

(ii) Obviously $\bar{R}$ is reflexive. Furthermore we have by $\mathsf{LSC}(t)$ for all $x, y, z \in \mathcal{X}$:

$$\begin{aligned} [\![x\bar{R}y \wedge_t y\bar{R}z]\!] &= (\sup_{i<\alpha}[\![xR_iy]\!])\, t\, (\sup_{i<\alpha}[\![yR_jz]\!]) \\ &= \sup_{i,j<\alpha} [\![xR_iy \wedge_t yR_jz]\!] \\ &\leq \sup_{i,j<\alpha} [\![xR_{\max\{i,j\}}z]\!] = [\![x\bar{R}z]\!] \end{aligned}$$

and hence the t-transitivity of $\bar{R}$. Again, if all fuzzy relations R_i are t-antisymmetric then the t-antisymmetry of $\bar{R}$ follows by corresponding calculations. QED

Proposition 2.32 *If R is a fuzzy preordering then the fuzzy relation*

$$Q =_{def} \{(x,y) \,\|\, xRy \wedge_t yRx\} \tag{2.36}$$

is a fuzzy equivalence relation.

Proof. Reflexivity and symmetry of Q are obvious by definition (2.36) and the reflexivity of R. And the t-transitivity of Q is a simple consequence of the t-transitivity of R. QED

Proposition 2.33 *Let R be a fuzzy preordering and define the fuzzy equivalence relation Q as in (2.36). Then by*

$$\hat{R} =_{def} \{(\langle a\rangle_Q, \langle b\rangle_Q) \,\|\, aRb\}$$

a fuzzy relation is defined in the quotient set $\mathcal{X}/Q$ which is a fuzzy partial ordering.

Proof. First we have to show that $\hat{R}$ is correctly defined, i.e. that this definition does not really depend on the elements a, b which describe the fuzzy equivalence classes $\langle a\rangle_Q, \langle b\rangle_Q$.

Hence consider a, a' such that $\langle a\rangle_Q = \langle a'\rangle_Q$. Then by $[\![a \,\varepsilon\, \langle a\rangle_Q]\!] = 1$ one has $[\![a \,\varepsilon\, \langle a'\rangle_Q]\!] = 1$, i.e. $[\![aQa']\!] = 1$. Hence immediately $[\![aRa']\!] = [\![a'Ra]\!] = 1$. Now we get for every x

$$[\![aRx]\!] = [\![a'Ra \wedge_t aRx]\!] \leq [\![a'Rx]\!] = [\![aRa' \wedge_t a'Rx]\!] \leq [\![aRx]\!]$$

by the t-transitivity of R and thus $[\![aRx]\!] = [\![a'Rx]\!]$. Therefore

$$[\![\langle a\rangle_Q \hat{R} \langle b\rangle_Q]\!] = [\![aRb]\!] = [\![a'Rb]\!] = [\![\langle a'\rangle_Q \hat{R} \langle b\rangle_Q]\!].$$

In the same way the independence of the choice of b is established.

The reflexivity of R is obvious. The t-transitivity results from the fact that for all fuzzy equivalence classes $\langle a\rangle_Q, \langle b\rangle_Q, \langle c\rangle_Q$:

$$\begin{aligned} &[\![\langle a\rangle_Q \hat{R} \langle b\rangle_Q \wedge_t \langle b\rangle_Q \hat{R} \langle c\rangle_Q]\!] \\ &\quad = [\![aRb \wedge_t bRc]\!] \leq [\![\langle a\rangle_Q \hat{R} \langle c\rangle_Q]\!]. \end{aligned}$$

Thus the antisymmetry of $\hat{R}$ remains. But immediately

$$[\![\langle a\rangle_Q \hat{R} \langle b\rangle_Q \wedge_t \langle b\rangle_Q \hat{R} \langle a\rangle_Q]\!] = [\![aRb \wedge_t bRa]\!] \leq [\![a \doteq b]\!]$$

and obviously

$$[\![a \doteq b]\!] \leq [\![\langle a\rangle_Q \doteq \langle b\rangle_Q]\!].$$

Thus $\hat{R}$ is a fuzzy partial ordering. QED

As a side remark let us remember that according to proposition 2.20 (iii) in case $\mathsf{LSC}(t)$ we have

$$\models \langle a\rangle_Q \equiv_t \langle b\rangle_Q \leftrightarrow_t aQb$$

i.e. in the present situation

$$\models \langle a\rangle_Q \equiv_t \langle b\rangle_Q \leftrightarrow_t aRb \wedge_t bRa.$$

If we assume that our universe of discourse $\mathcal{X}$ is already a class of fuzzy sets, which means that our fuzzy subsets of $\mathcal{X}$ and our fuzzy relations in $\mathcal{X}$ are fuzzy sets of higher level in the terminology of GOTTWALD (1979), then it seems reasonable to discuss another, suitably changed version of antisymmetry of such a fuzzy relation R characterized through the definition

$$\models \forall x \forall y (xRy \wedge_t yRx \rightarrow_t x \equiv_t y).$$

If in that case we repeat the construction of $\hat{R}$ from proposition 2.33 we find that this fuzzy relation $\hat{R}$ is antisymmetric in this new sense too: we have

$$[\![\langle a\rangle_Q \hat{R} \langle b\rangle_Q \wedge_t \langle b\rangle_Q \hat{R} \langle a\rangle_Q]\!] = [\![aRb \wedge_t bRa]\!] \leq [\![\langle a\rangle_Q \equiv_t \langle b\rangle_Q]\!]$$

and hence

$$\models \forall x \forall y (\langle x\rangle_Q \hat{R} \langle y\rangle_Q \wedge_t \langle y\rangle_Q \hat{R} \langle x\rangle_Q \rightarrow_t \langle x\rangle_Q \equiv_t \langle y\rangle_Q).$$

As our last topic in this section we discuss a fuzzification of the well-known SZPILRAJN's theorem on the extension of a partial ordering to a linear ordering. We get here a generalization of the fuzzy version of this theorem as proved in ZADEH (1971).

Definition 2.17 *Given any t-conorm s_{t_1} we denote a fuzzy relation R as* s_{t_1}-linear *iff*

$$\models \forall x \forall y (xRy \vee_{t_1} yRx),$$

and we denote R as weakly linear *iff $supp(R)$ is a linear relation in the usual sense. Additionally we call R a* s_{t_1}-linear (weakly linear) fuzzy ordering *iff R is a fuzzy partial ordering and also s_{t_1}-linear (weakly linear).*

Proposition 2.34 *Suppose that the t-norm t does not have zero divisors. Then for every fuzzy partial ordering R there exists a s_{t_1}-linear fuzzy ordering S_R (s_{t_1} any t-conorm) such that $\models R \subseteqq_t S_R$. If the fuzzy partial ordering R has the further property that there exists $u \in (0,1]$ such that for all $a, b \in \mathcal{X}$:*

$$(a,b) \in \text{supp}(R) \Rightarrow [\![aRb]\!] \geq u,$$

then there exists a weakly linear fuzzy ordering $\widetilde{S_R}$ such that $\models R \subseteqq_t \tilde{S}_R$ and for all $a, b \in \mathcal{X}$ also

$$(a,b) \in \text{supp}(R) \Rightarrow [\![aRb]\!] = [\![a\widetilde{S_R}b]\!].$$

Proof. By proposition 2.30 under the present assumptions supp(R) is a partial ordering (in the usual sense). By the classical SZPILRAJN's theorem hence there exists an usual linear ordering S_R^* which extends supp(R). Define the fuzzy relation S_R by

$$[\![(x,y) \,\varepsilon\, S_R]\!] = \begin{cases} 1 & \text{iff} \quad xS_R^*y \\ 0 & \text{otherwise.} \end{cases}$$

Obviously S_R is a fuzzy partial ordering which furthermore is s_{t_1}-linear for each t-conorm s_{t_1} because of the usual linearity of S_R^*. By the way, the construction of S_R guarantees $S_R^* = \text{supp}(S_R)$.

Of course, by the construction of S_R we get

$$\models R \subseteqq S_R$$

simply from supp$(R) \subseteq S_R^* = \text{supp}(S_R)$. Yet, again by this construction, all the ordered pairs $(a,b) \in \text{supp}(R)$ which belong to R with a membership degree different from 1 have a strictly greater membership degree in S_R than in R.

Assuming the existence of some $u \in (0,1]$ such that

$$(a,b) \in \text{supp}(R) \Rightarrow [\![aRb]\!] \geq u \tag{2.37}$$

for all $a, b \in \mathcal{X}$, we define an auxiliary fuzzy relation R_u by

$$[\![(x,y) \varepsilon R_u]\!] = \max\{[\![(x,y) \varepsilon R]\!], u\}$$

and put

$$\widetilde{S_R} = S_R \cap_t R_u$$

(which gives the same fuzzy relation $\widetilde{S_R}$ for each t-norm t by construction of S_R). Obviously again

$$\models R \subseteqq \widetilde{S_R}.$$

But now additionally

$$(a,b) \in \operatorname{supp}(R) \Rightarrow [\![aRb]\!] = [\![a\widetilde{S_R}b]\!];$$

and again $\widetilde{S_R}$ is a fuzzy partial ordering which now, yet, is weakly linear by the linearity of S_R^*. In order to prove the t-transitivity and t-antisymmetry of $\widetilde{S_R}$ we essentially have to use (2.37) and the fact that

$$(a,b) \in \operatorname{supp}(\widetilde{S_R}) \setminus \operatorname{supp}(R) \Rightarrow [\![a\widetilde{S_R}b]\!] = u.$$

The details again are a routine matter. QED

2.5 Graded properties of fuzzy relations

The properties of fuzzy relations which were introduced in the previous definition 2.10 are crisp in the sense that a fuzzy relation R either has one of those properties or does not have that property. Therefore the situation here is comparable with that we had before with the ZADEH's crisp inclusion (2.4) for fuzzy sets. And as we in that case were interested in generalizing that inclusion $\subset$ to a graded one by introducing the fuzzified inclusions $\subseteqq_t$ in definition 2.3, we now again look for possibilities to generalize the relation properties of definition 2.10 in such a way that fuzzified, i.e. graded properties result.

The fact that we always intended to write down our notions in the language of (a suitable) many-valued logic now opens the door to get such graded properties in a canonical manner out of those introduced in definition 2.10. The crucial point is that definition 2.10 used the logical validity of suitable well-formed formulas to define the relation properties. Therefore deleting this demand of logical validity and retaining the characteristic well-formed formulas immediately opens a way to get graded properties of the type we are looking for.

In the formal sense these graded properties themselves are to be represented by many-valued predicates of our language of many-valued logic.

Definition 2.18 *For each fuzzy relation R and any t-norm t with property* $\mathsf{LSC}(t)$ *let*

$$\begin{aligned}
Refl(R) &=_{def} \forall x(xRx), \\
Irrefl_t(R) &=_{def} \forall x -_t (xRx), \\
Trans_t(R) &=_{def} \forall x \forall y \forall z(xRy \wedge_t yRz \rightarrow_t xRz), \\
Symm_t(R) &=_{def} \forall x \forall y(xRy \rightarrow_t yRx), \\
Antisymm_t(R) &=_{def} \forall x \forall y(xRy \wedge_t yRx \rightarrow_t x \doteq y), \\
Asymm_t(R) &=_{def} \forall x \forall y -_t (xRy \wedge_t yRx).
\end{aligned}$$

To get a simpler approach, in comparison with definition 2.10, we here dispense with the use of a separate negation function n to define the predicates $Irrefl_t$ and $Asymm_t$. On the other hand this change from crisp to graded properties also causes a dependency on the involved t-norm for the properties of irreflexivity and symmetry.

Immediately one has e.g. the result that

$$\models Trans_t(R) \quad \Leftrightarrow \quad R \ t\text{-transitive}$$

and the corresponding results for all the other graded properties with respect to their crisp versions

As in the case of the crisp versions of relation properties, a characterization of these graded properties with genuinely set theoretic notions is now also possible.

Proposition 2.35 *For each fuzzy relation R it holds true that*

$$\begin{aligned}
&\models \quad Refl(R) \leftrightarrow \Delta_{\mathcal{X}} \subseteqq R, \\
&\models \quad Irrefl_t(R) \leftrightarrow R \cap_t \Delta_{\mathcal{X}} \equiv_t \emptyset, \\
&\models \quad Trans_t(R) \leftrightarrow R \circ_t R \subseteqq_t R, \\
&\models \quad Symm_t(R) \leftrightarrow R \subseteqq_t R^{-1}, \\
&\models \quad Antisymm_t(R) \leftrightarrow R \cap_t R^{-1} \subseteqq_t \Delta_{\mathcal{X}}, \\
&\models \quad Asymm_t(R) \leftrightarrow R \cap_t R^{-1} \equiv_t \emptyset.
\end{aligned}$$

Proof. Straightforward from the respective definitions, additionally using proposition 2.5 (i) for the cases of the predicates $Irrefl_t$ and $Asymm_t$. QED

In general, this "fuzzification" of properties of fuzzy relations is nothing completely new. Before we essentially did the same with our definition 2.3 which introduced graded versions $\subseteq_t$ and $\equiv_t$ of inclusion and equality for fuzzy sets. The small difference is only that there the generalization to graded relations concerned binary relations, i.e. binary predicates for fuzzy sets, and now we are concerned with generalized properties, i.e. unary predicates for fuzzy relations.

As a relatively simple and very well-known example of a not completely elementary special type of fuzzy relations, let us further on take a look at fuzzy ordering relations which – for the crisp case – we already discussed in the previous section. Now a graded notion of fuzzy partial ordering shall be introduced. First we focus on the reflexive case.

Definition 2.19 *Let $R \in \mathbb{F}(\mathcal{X} \times \mathcal{X})$ be a fuzzy relation and t any t-norm with property* $\mathsf{LSC}(t)$. *Put*

$$FPO_t(R) =_{def} Refl(R) \wedge_t Trans_t(R) \wedge_t Antisymm_t(R).$$

Obviously by $\mathsf{LSC}(t)$ and basic properties of t-norms one has

$$\models FPO_t(R) \quad \Leftrightarrow \quad R \text{ fuzzy partial ordering.}$$

Corollary 2.36 *Let $R \in \mathbb{F}(\mathcal{X} \times \mathcal{X})$ be a fuzzy relation and t any t-norm with property* $\mathsf{LSC}(t)$. *Then*

$$\models FPO_t(R) \leftrightarrow \Delta_{\mathcal{X}} \subseteq_t R \wedge_t R \circ_t R \subseteq_t R \wedge_t R \cap_t R^{-1} \equiv_t \emptyset.$$

Proof. Obvious by proposition 2.35.

Let us first consider the usual duality between reflexive and irreflexive fuzzy ordering relations established by changing from R to $R^+ = R \cup_t \Delta_{\mathcal{X}}$ and from R to $R^- = R \setminus_t \Delta_{\mathcal{X}} = \{(x, y) \parallel (x, y) \varepsilon R \wedge_t -_t(x, y) \varepsilon \Delta_{\mathcal{X}}\}$, i.e. to $R^- = R \cap_t \complement\Delta_{\mathcal{X}}$, cf. definition 2.15 and proposition 2.27.

Proposition 2.37 *Let $R \in \mathbb{F}(\mathcal{X} \times \mathcal{X})$ be a fuzzy relation and t any t-norm with the property* $\mathsf{LSC}(t)$. *Then there hold true*

$$\begin{array}{ll} (i) & \models Refl(R \cup_t \Delta_{\mathcal{X}}), \\ (ii) & \models Irrefl_t(R \setminus_t \Delta_{\mathcal{X}}), \\ (iii) & \models Trans_t(R) \to_t Trans_t(R \cup_t \Delta_{\mathcal{X}}), \\ (iv) & \models Trans_t(R \setminus_t \Delta_{\mathcal{X}}) \to_t Trans_t(R). \end{array}$$

Proof. (i) and (ii) are obvious. For (iii) one has to prove the inequality

$$[\![\forall x, y, z(xRy \wedge_t yRz \to_t xRz)]\!] \leq \\ [\![\forall x, y, z(xR^+y \wedge_t yR^+z \to_t xR^+z)]\!]$$

for $R^+ = R \cup_t \Delta_{\mathcal{X}}$. To get this we first show the inequality

$$[\![aRb \wedge_t bRc \to_t aRc]\!] \leq [\![aR^+b \wedge_t bR^+c \to_t aR^+c]\!] \tag{2.38}$$

for arbitrary $a, b, c \in \mathcal{X}$.

In the case that $a = b$ holds true it remains to be shown that

$$[\![aRa \wedge_t aRc \to_t aRc]\!] \leq [\![aR^+a \wedge_t aR^+c \to_t aR^+c]\!].$$

But this obviously holds true because both these implications have truth degree 1: in each of them the truth degree of the antecedent is not greater than the truth degree of the succedent.

In the case that $b = c$ holds true, essentially the same argument gives (2.38).

Thus the case $a \neq b \neq c$ remains to be considered. Now, however, one has

$$[\![aRb]\!] = [\![aR^+b]\!], \qquad [\![bRc]\!] = [\![bR^+c]\!], \qquad [\![aRc]\!] \leq [\![aR^+c]\!]$$

and thus again (2.38).

Therefore (2.38) generally holds true. But then obviously also

$$[\![\forall x, y, z(xRy \wedge_t yRz \to_t xRz)]\!] \leq [\![aR^+b \wedge_t bR^+c \to_t aR^+c]\!]$$

holds true. Here the left-hand side of the inequality is independent of the choice of a, b, c. That means that even the infimum (with respect to a, b, c) of the right-hand side is not smaller than the left-hand side, but this is exactly (iii) according to definition 2.18.

Finally (iv) follows along the same line of argument as (iii). QED

Unfortunately the transitivity property cannot be transfered from R to $R \setminus_t \Delta_{\mathcal{X}}$ in a manner directly analogous to (iii). However, that transfer becomes possible with an additional assumption.

Proposition 2.38 *Let $R \in \mathbb{F}(\mathcal{X} \times \mathcal{X})$ be a fuzzy relation and $\mathbf{t}$ any t-norm with the property* $\mathsf{LSC}(\mathbf{t})$. *Then one has*

$$\models \mathit{Trans}_{\mathbf{t}}(R) \wedge_{\mathbf{t}} \mathit{Antisymm}_{\mathbf{t}}(R) \to \mathit{Trans}_{\mathbf{t}}(R \setminus_{\mathbf{t}} \Delta_{\mathcal{X}}).$$

Proof. As in part (iii) of the last proof it is enough to show that for $R^- = R \setminus_{\mathbf{t}} \Delta_{\mathcal{X}} = R \cap_{\mathbf{t}} \mathbf{C}\Delta_{\mathcal{X}}$ and all $a, b, c \in \mathcal{X}$

$$[\![\mathit{Trans}_{\mathbf{t}}(R) \wedge_{\mathbf{t}} \mathit{Antisymm}_{\mathbf{t}}(R)]\!] \leq [\![aR^-b \wedge_{\mathbf{t}} bR^-c \to_{\mathbf{t}} aR^-c]\!]. \quad (2.39)$$

If $c \neq a$ holds true, then $[\![aR^-c]\!] = [\![aRc]\!]$. By $\models R^- \subseteqq_{\mathbf{t}} R$ hence

$$[\![aRb \wedge_{\mathbf{t}} bRc \to_{\mathbf{t}} aRc]\!] \leq [\![aR^-b \wedge_{\mathbf{t}} bR^-c \to_{\mathbf{t}} aRc]\!]$$

and thus even

$$[\![\mathit{Trans}_{\mathbf{t}}(R)]\!] \leq [\![aR^-b \wedge_{\mathbf{t}} bR^-c \to_{\mathbf{t}} aRc]\!]$$

holds true. So (2.39) is established in this case.

Now assume $c = a$. Then $[\![aR^-c]\!] = 0$ and (2.39) becomes

$$[\![\mathit{Trans}_{\mathbf{t}}(R) \wedge_{\mathbf{t}} \mathit{Antisymm}_{\mathbf{t}}(R)]\!] \leq [\![-_{\mathbf{t}}(aR^-b \wedge_{\mathbf{t}} bR^-a)]\!]. \quad (2.40)$$

In the case that $a = b$ one has $[\![aR^-b]\!] = 0$ and thus $[\![-_{\mathbf{t}}(aR^-b \wedge_{\mathbf{t}} bR^-a)]\!] = 1$, which immediately gives (2.40). And in the case that $a \neq b$ one has $[\![a \doteq b]\!] = 0$ and thus

$$[\![-_{\mathbf{t}}(aR^-b \wedge_{\mathbf{t}} bR^-a)]\!] = [\![aR^-b \wedge_{\mathbf{t}} bR^-a \to_{\mathbf{t}} a \doteq b]\!],$$

hence

$$[\![\mathit{Antisymm}_{\mathbf{t}}(R)]\!] \leq [\![aR^-b \wedge_{\mathbf{t}} bR^-a \to_{\mathbf{t}} a \doteq b]\!]$$

and therefore (2.40). Thus (2.39) holds true in any case. QED

Similar transfer results hold true regarding antisymmetry and asymmetry.

Proposition 2.39 *Let $R \in \mathbb{F}(\mathcal{X} \times \mathcal{X})$ be a fuzzy relation and t any t-norm with property* LSC(t). *Then there hold true*

$$\begin{array}{ll} (i) & \models Antisymm_t(R) \rightarrow Asymm_t(R \setminus_t \Delta_{\mathcal{X}}), \\ (ii) & \models Asymm_t(R \setminus_t \Delta_{\mathcal{X}}) \leftrightarrow Antisymm_t(R \setminus_t \Delta_{\mathcal{X}}), \\ (iii) & \models Asymm_t(R) \rightarrow Antisymm_t(R \cup_t \Delta_{\mathcal{X}}). \end{array}$$

Proof. The same type of estimations as in the proofs of the last two propositions yield the results. QED

By and large, the basic links for a generalization of the usual duality of reflexive and irreflexive partial orderings seem to be established with those results. What remains to be given is a definition of graded irreflexive fuzzy partial orderings. Caused by the process of grading this definition is not as obvious as in the reflexive case. The background difficulty is the well-known fact for classical irreflexive partial orderings that they can either be defined by the requirements of irreflexivity and transitivity or – equivalently – by those of irreflexivity, transitivity, and asymmetry.

In the present situation therefore we are first interested whether transitivity together with irreflexivity also imply asymmetry in our more general context. Indeed we have

Proposition 2.40 *Let $R \in \mathbb{F}(\mathcal{X} \times \mathcal{X})$ be a fuzzy relation and t any t-norm with property* LSC(t). *Then it holds true that*

$$\models Trans_t(R) \wedge_t Irrefl_t(R) \rightarrow Asymm_t(R).$$

Proof. As in the foregoing proofs it is enough to prove for any $a, b \in \mathcal{X}$

$$\models -_t aRa \wedge_t \forall z(aRb \wedge_t bRz \rightarrow_t aRz) \rightarrow_t (aRb \rightarrow_t -_t bRa). \quad (2.41)$$

Using proposition 1.19 (i) twice and also proposition 1.27 (i) one finds that (2.41) is equivalent to

$$\models -_t aRa \wedge_t aRb \wedge_t (aRb \rightarrow_t \forall z(bRz \rightarrow_t aRz)) \rightarrow_t -_t bRa.$$

But now using propositions 1.17 (i) and 1.23 (iv) together with the antimonotonicity of the Φ-operators in the first argument one has for the truth degree of this expression

$$\begin{aligned}
[\![-_t aRa \wedge_t aRb \wedge_t (aRb \rightarrow_t \forall z(bRz \rightarrow_t aRz)) \rightarrow_t -_t bRa]\!]& \\
\geq [\![-_t aRa \wedge_t \forall z(bRz \rightarrow_t aRz) \rightarrow_t -_t bRa]\!]& \\
= [\![\forall z(bRz \rightarrow_t aRz) \rightarrow_t (-_t aRa \rightarrow_t -_t bRa)]\!]& \\
\geq [\![\forall z(bRz \rightarrow_t aRz) \rightarrow_t (bRa \rightarrow_t aRa)]\!]& \\
= 1&
\end{aligned}$$

and thus (2.41). QED

Therefore we decide to base irreflexive partial orderings only on transitivity and irreflexivity.

Definition 2.20 *Let $R \in \mathbb{F}(\mathcal{X} \times \mathcal{X})$ be a fuzzy relation and $\mathbf{t}$ any t-norm with property* $\mathsf{LSC}(\mathbf{t})$. *Then we put*

$$FPO^*_{\mathbf{t}}(R) =_{def} Trans_{\mathbf{t}}(R) \wedge_{\mathbf{t}} Irrefl_{\mathbf{t}}(R).$$

With this definition all the notions and preliminary results are completed which we need to state the generalized connection between the reflexive and irreflexive graded fuzzy partial ordering relations.

Proposition 2.41 *Let $R \in \mathbb{F}(\mathcal{X} \times \mathcal{X})$ be a fuzzy relation and $\mathbf{t}$ any t-norm with property* $\mathsf{LSC}(\mathbf{t})$. *Then there hold true*

$$\begin{aligned}
&(i) && \models Trans_{\mathbf{t}}(R) \wedge_{\mathbf{t}} Antisymm_{\mathbf{t}}(R) \rightarrow FPO^*_{\mathbf{t}}(R \setminus_{\mathbf{t}} \Delta_{\mathcal{X}}), \\
&(ii) && \models FPO_{\mathbf{t}}(R) \rightarrow FPO^*_{\mathbf{t}}(R \setminus_{\mathbf{t}} \Delta_{\mathcal{X}}), \\
&(iii) && \models FPO^*_{\mathbf{t}}(R) \wedge_{\mathbf{t}} Asymm_{\mathbf{t}}(R) \rightarrow FPO_{\mathbf{t}}(R \cup_{\mathbf{t}} \Delta_{\mathcal{X}}).
\end{aligned}$$

Proof. Because of proposition 2.37 (ii) claim (i) is a direct consequence of proposition 2.38. Of course, (ii) is a simple corollary of (i). And (iii) results from Proposition 2.37 (i) together with proposition 2.37 (iii) and proposition 2.39 (iii). QED

The last claim of this proposition could be read as indicating that our definition 2.20 was the wrong choice. But considering instead of $FPO^*_{\mathbf{t}}$ the many-valued predicate

$$FPO^{**}_{\mathbf{t}}(R) =_{def} Trans_{\mathbf{t}}(R) \wedge_{\mathbf{t}} Irrefl_{\mathbf{t}}(R) \wedge_{\mathbf{t}} Asymm_{\mathbf{t}}(R) \qquad (2.42)$$

one indeed would be able to prove

$$\models FPO_t^{**}(R) \rightarrow FPO_t(R \cup_t \Delta_{\mathcal{X}}),$$

yet instead of proposition 2.41 (ii) then only

$$\models FPO_t(R) \wedge_t Antisymm_t(R) \rightarrow FPO_t^{**}(R \setminus_t \Delta_{\mathcal{X}})$$

would be provable.

All together that means that neither definition 2.20 nor (2.42) gives the simple results one has for crisp partial orderings. Only a restriction of the t-norm under consideration allows for a simplification of proposition 2.41 (ii).

Corollary 2.42 *For each fuzzy relation* $R \in \mathbb{F}(\mathcal{X} \times \mathcal{X})$ *and the t-norm* $t = t_G = \min$ *one has*

$$\models FPO_{t_G}^{*}(R) \rightarrow FPO_{t_G}(R \cup \Delta_{\mathcal{X}}).$$

Proof. In this special case one has

$$\begin{aligned} [\![FPO_{t_G}^{*}(R)]\!] &= [\![FPO_{t_G}^{*}(R) \wedge FPO_{t_G}^{*}(R)]\!] \\ &\leq [\![FPO_{t_G}^{*}(R) \wedge Asymm_{t_G}(R)]\!] \end{aligned}$$

and thus the result immediately follows from proposition 2.41 (ii). QED

Besides the ordering relations the equivalence relations are of fundamental importance in crisp mathematics. Like ordering relations we already introduced fuzzy equivalence relations in the last section, cf. definition 2.11. Here we intend to have a graded property of being an equivalence relation.

Definition 2.21 *Let* $R \in \mathbb{F}(\mathcal{X} \times \mathcal{X})$ *be a fuzzy relation and* t *any t-norm with property* $\mathsf{LSC}(t)$. *Then we define*

$$Eqrel_t(R) =_{def} Refl(R) \wedge_t Symm_t(R) \wedge_t Trans_t(R).$$

We do not intend to present here an extended theory of graded fuzzy equivalence relations. That is to the same degree a future research requirement as a more elaborated theory of graded fuzzy orderings. But we shall consider one topic, the fuzzification of the well-known fact that any two distinct equivalence classes of a (crisp) equivalence relation are disjoint sets. There was a first type of fuzzification, for fuzzy relations which "really", i.e.

to truth degree 1, are fuzzy equivalence relations, presented in proposition 2.20 (iv), a result that can equivalently be written as

$$\models \langle a \rangle_R \cap_t \langle b \rangle_R \equiv_t \emptyset \;\leftrightarrow\; -_t aRb$$

or as

$$\models \langle a \rangle_R \cap_t \langle b \rangle_R \not\equiv_t \emptyset \;\leftrightarrow\; aRb.$$

Further partial fuzzifications of this result read as follows.

Proposition 2.43 *Let $R \in \mathbb{F}(\mathcal{X} \times \mathcal{X})$ be a fuzzy relation and t any t-norm with property $\mathsf{LSC}(t)$. Then there hold true for any $a, b \in \mathcal{X}$*

$$\begin{array}{ll} (i) & \models Trans_t(R) \to (aRb \to_t \langle a \rangle_R \subseteq_t \langle b \rangle_R), \\ (ii) & \models Refl(R) \to ((\langle a \rangle_R \subseteq_t \langle b \rangle_R \to_t aRb), \\ (iii) & \models Trans_t(R) \wedge_t Symm_t(R) \to \\ & \qquad\qquad ((\langle a \rangle_R \cap_t \langle b \rangle_R \not\equiv_t \emptyset \to_t aRb). \end{array}$$

Proof. (i) By the corresponding definitions and propositions 1.19 (i) and 1.17 (i), one has the following inequalities for truth degrees:

$$\begin{aligned} [\![Trans_t(R) \wedge_t aRb]\!] &= [\![\forall x, y, z(xRy \wedge_t yRz \to_t xRz) \wedge_t aRb]\!] \\ &\leq [\![\forall x(aRb \wedge_t (xRa \wedge_t aRb \to_t xRb)]\!] \\ &= [\![\forall x(aRb \wedge_t)(aRbto_t(xRa \to_t xRb)))]\!] \\ &\leq [\![\forall x(xRa \to_t xRb))]\!] \\ &= [\![\langle a \rangle_R \subseteq_t \langle b \rangle_R]\!]. \end{aligned}$$

Hence (i) immediately follows again using proposition 1.19 (i).

(ii) With essentially the same logical background as in the foregoing proof of claim (i) one gets (ii) by using proposition 1.19 (i) together with

$$\begin{aligned} [\![Refl(R) \wedge_t \langle a \rangle_R \subseteq_t \langle b \rangle_R]\!] &\leq [\![aRa \wedge_t (aRa \to_t aRb)]\!] \\ &\leq [\![aRb]\!]. \end{aligned}$$

(iii) According to proposition 2.5 (ii) one has

$$\begin{aligned} [\![\langle a \rangle_R \cap_t \langle b \rangle_R \not\equiv_t \emptyset]\!] &= [\![\exists x(x \,\varepsilon\, \langle a \rangle_R \cap_t \langle b \rangle_R)]\!] \\ &= [\![\exists x(xRa \wedge_t xRb)]\!]. \end{aligned}$$

Therefore with propositions 1.28 (i) and 1.17 (i) one gets as an estimation of truth degrees:

$$\begin{aligned}
&[\![Trans_t(R) \wedge_t Symm_t(R) \wedge_t \exists x(xRa \wedge_t xRb)]\!] \\
&\quad = [\![\exists x(Trans_t(R) \wedge_t Symm_t(R) \wedge_t xRa \wedge_t xRb)]\!] \\
&\quad \leq [\![\exists x(Trans_t(R) \wedge_t (xRa \rightarrow_t aRx) \wedge_t xRa \wedge_t xRb)]\!] \\
&\quad \leq [\![\exists x(Trans_t(R) \wedge_t aRx \wedge_t xRb)]\!] \\
&\quad \leq [\![\exists x((aRx \wedge_t xRb \rightarrow_t aRb) \wedge_t aRx \wedge_t xRb)]\!] \\
&\quad \leq [\![\exists x(aRb)]\!] \ = \ [\![aRb]\!].
\end{aligned}$$

Now once again the application of proposition 1.19 (i) gives the result. QED

Corollary 2.44 *Under the same conditions as in proposition 2.42 one has*

$$\begin{array}{ll}
(i) & \models Refl(R) \wedge Trans_t(R) \rightarrow ((\langle a\rangle_R \subseteq_t \langle b\rangle_R \leftrightarrow_t aRb), \\
(ii) & \models Trans_t(R) \wedge_t Symm_t(R) \wedge_t -_t aRb \rightarrow \\
 & \qquad\qquad\qquad \langle a\rangle_R \cap_t \langle b\rangle_R \equiv_t \emptyset, \\
(iii) & \models [Trans_t(R)]^2 \rightarrow (aRb \wedge_t bRa \rightarrow_t \langle a\rangle_R \equiv_t \langle b\rangle_R), \\
(iv) & \models Trans_t(R) \rightarrow \\
 & \qquad (aRb \wedge bRa \rightarrow_t \langle a\rangle_R \subseteq_t \langle b\rangle_R \wedge \langle b\rangle_R \subseteq_t \langle a\rangle_R).
\end{array}$$

Proof. (i) results via proposition 1.19 (iv) from proposition 2.43 (i) and (ii) using the fact that $[\![H_1 \leftrightarrow_t H_2]\!] = [\![(H_1 \rightarrow_t H_2) \wedge (H_2 \rightarrow_t H_1)]\!]$ always holds true.

(ii) We start from the succedent of the implication in the previous proposition 2.43 (iii) and have by propositions 1.22 (i) and 1.19 (ii)

$$\begin{aligned}
&\models ((\langle a\rangle_R \cap_t \langle b\rangle_R \not\equiv_t \emptyset \rightarrow_t aRb) \\
&\qquad \rightarrow (-_t((\langle a\rangle_R \cap_t \langle b\rangle_R \equiv_t \emptyset) \rightarrow_t -_t -_t aRb)
\end{aligned}$$

and by proposition 1.23 (iii) also

$$\begin{aligned}
&\models (-_t((\langle a\rangle_R \cap_t \langle b\rangle_R \equiv_t \emptyset) \rightarrow_t -_t -_t aRb) \\
&\qquad \rightarrow (-_t aRb \rightarrow_t -_t -_t ((\langle a\rangle_R \cap_t \langle b\rangle_R \equiv_t \emptyset)).
\end{aligned}$$

Therefore from propositions 2.5 (ii) and 2.43 (iii) by an iterated application of (the consequence rule corresponding to) proposition 1.19 (ii) one finds

$$\models Trans_t(R) \wedge_t Symm_t(R) \rightarrow -_t -_t -_t \exists x(x \,\varepsilon\, \langle a\rangle_R \cap_t \langle b\rangle_R).$$

With proposition 1.22 (ii) that means

$$\models Trans_t(R) \wedge_t Symm_t(R) \rightarrow -_t\exists x(x \varepsilon \langle a\rangle_R \cap_t \langle b\rangle_R)$$

which gives the final

$$\models Trans_t(R) \wedge_t Symm_t(R) \rightarrow \langle a\rangle_R \cap_t \langle b\rangle_R \equiv_t \emptyset$$

by again applying proposition 2.5 (ii).

(iii) Let $H(a,b)$ be the formula $aRb \rightarrow_t \langle a\rangle_R \subseteq_t \langle b\rangle_R$. Taking the formula $Trans_t(R) \rightarrow H(a,b)$, whose logical validity is stated in proposition 2.43 (i), together with the formula $Trans_t(R) \rightarrow H(b,a)$ which results from the earlier one by interchanging a and b, that gives via proposition 1.19 (iii)

$$\models [Trans_t(R)]^2 \rightarrow (H(a,b) \wedge_t H(b,a)). \tag{2.43}$$

Now both formulas $H(a,b)$ and $H(b,a)$ themselves are implications. Thus again applying proposition 1.19 (iii) to the succedent of (2.43) finally gives our claim (iii) according to the definition of $\equiv_t$.

(iv) follows by the same arguments, but refering to proposition 1.19 (iv) instead of 1.19 (iii) and thus having

$$\models Trans_t(R) \rightarrow (H(a,b) \wedge H(b,a))$$

instead of (2.43). QED

For the t-norm $t = t_G = \min$ even a little more becomes provable.

Corollary 2.45 *For each fuzzy relation $R \in I\!F(\mathcal{X} \times \mathcal{X})$ and all $a,b \in \mathcal{X}$ there hold true*

(i) $\models Trans_{t_G}(R) \rightarrow (aRb \rightarrow_{t_G} \langle a\rangle_R \equiv_{t_G} \langle b\rangle_R)),$

(ii) $\models Trans_{t_G}(R) \wedge Symm_{t_G} \rightarrow$
$$((\langle a\rangle_R \cap \langle b\rangle_R \equiv_{t_G} \langle a\rangle_R \cap \{x \,\|\, aRb\}).$$

Proof. (i) results immediately from corollary 2.44 (ii) by the idempotency of the t-norm t_G.

(ii) With regard to proposition 2.43 (iii) we have the following sequence of transformations of the truth degree of the respective succedent:

$$\begin{aligned}
&[\![\langle a\rangle_R \cap \langle b\rangle_R \not\equiv_{t_G} \emptyset \rightarrow_{t_G} aRb]\!] \\
&\quad = [\![\exists x(x \varepsilon \langle a\rangle_R \cap \langle b\rangle_R) \rightarrow_{t_G} aRb]\!] \\
&\quad = [\![\forall x(x \varepsilon \langle a\rangle_R \wedge \langle b\rangle_R \rightarrow_{t_G} aRb)]\!] \\
&\quad = [\![\forall x(x \varepsilon \langle a\rangle_R \wedge \langle b\rangle_R \rightarrow_{t_G} aRb \wedge x \varepsilon \langle a\rangle_R)]\!]
\end{aligned}$$

whose last transformation rests on simple properties of the min-operator t_G and its corresponding Φ-operator φ_G.

Using the fact that for each x

$$[\![aRb]\!] = [\![x \,\varepsilon\, \{y \,\|\, aRb\}]\!] \tag{2.44}$$

and changing the bound variable y of the generalized class term $\{y \,\|\, aRb\}$ into x gives

$$[\![\langle a\rangle_R \cap \langle b\rangle_R \not\equiv_{t_G} \emptyset \rightarrow_{t_G} aRb]\!] = [\![\langle a\rangle_R \cap \langle b\rangle_R \subseteqq_{t_G} \langle a\rangle_R \cap \{x \,\|\, aRb\}]\!]$$

and thus all together

$$\models Trans_{t_G}(R) \wedge Symm_{t_G} \rightarrow \\ (\langle a\rangle_R \cap \langle b\rangle_R \subseteqq_{t_G} \langle a\rangle_R \cap \{x \,\|\, aRb\}). \tag{2.45}$$

On the other hand starting from proposition 2.43 (i) we have the following sequence of transformations of the truth degree of the respective succedent:

$$\begin{aligned} &[\![aRb \rightarrow_{t_G} \langle a\rangle_R \subseteqq_{t_G} \langle b\rangle_R]\!] \\ &\qquad = [\![aRb \rightarrow_{t_G} \forall x(x \,\varepsilon\, \langle a\rangle_R \rightarrow_{t_G} x \,\varepsilon\, \langle b\rangle_R)]\!] \\ &\qquad = [\![\forall x(aRb \wedge x \,\varepsilon\, \langle a\rangle_R \rightarrow_{t_G} x \,\varepsilon\, \langle b\rangle_R)]\!] \\ &\qquad = [\![\forall x(aRb \wedge x \,\varepsilon\, \langle a\rangle_R \rightarrow_{t_G} x \,\varepsilon\, \langle a\rangle_R \wedge x \,\varepsilon\, \langle b\rangle_R)]\!] \end{aligned}$$

and thus, again using (2.44),

$$\models Trans_{t_G}(R) \rightarrow (\langle a\rangle_R \cap \{x \,\|\, aRb\} \subseteqq_{t_G} \langle a\rangle_R \cap \langle b\rangle_R). \tag{2.46}$$

Combining (2.45) and (2.46) via proposition 1.19 (iv) then gives the result.

QED

Having for a fuzzy relation treated their property to be a fuzzy partial ordering relation, as well as their property to be a fuzzy equivalence relation both as graded predicates, it remains, with respect to the topics discussed in the previous section, to take a look at the fuzzified transitive hull of a fuzzy relation and to see which results on those t-transitive hulls were not given in a "truly fuzzified" form there. This obviously concerns proposition 2.23 and proposition 2.22 (ii).

Therefore let us try to generalize the last mentioned result too. We immediately have

$$\models R \equiv_t \mathcal{T}_t(R) \leftrightarrow \mathcal{T}_t(R) \subseteqq_t R$$

because of proposition 2.22 (i). Furthermore we have already proved the results

$$\models Refl(R) \rightarrow \Delta_{\mathcal{X}} \subseteqq_{t} R,$$
$$\models \top \rightarrow R \subseteqq_{t} R,$$
$$\models Trans_{t}(R) \rightarrow R^2 \subseteqq_{t} R$$

the second of which, however, was previously given as $\models R \subseteqq_{t} R$. Thus we have according to proposition 1.19 (iv)

$$\models Refl(R) \wedge \top \wedge Trans_{t}(R) \rightarrow (\Delta_{\mathcal{X}} \subseteqq_{t} R \wedge R \subseteqq_{t} R \wedge R^2 \subseteqq_{t} R)$$

and hence because of proposition 2.3 (i)

$$\models Refl(R) \wedge Trans_{t}(R) \rightarrow \bigcup_{i=0}^{2} R^i \subseteqq_{t} R. \tag{2.47}$$

From $R^3 = R^2 \circ_{t} R$ according to definition 2.13 together with proposition 2.15 (i), i.e. the monotonicity of $\circ_{t}$, we find

$$\models Trans_{t}(R) \rightarrow R^3 \subseteqq_{t} R^2$$

and thus

$$\models [Trans_{t}(R)]^2 \rightarrow R^3 \subseteqq_{t} R. \tag{2.48}$$

Unfortunately this antecedent $[Trans_{t}(R)]^2$ cannot be simplified in such a way that (2.48) holds true with $Trans_{t}(R)$ instead of $[Trans_{t}(R)]^2$. Thus we only get from (2.47) and (2.48)

$$\models Refl(R) \wedge [Trans_{t}(R)]^2 \rightarrow \bigcup_{i=0}^{3} R^i \subseteqq_{t} R. \tag{2.49}$$

using proposition 1.19 (iv) and the fact that

$$[\![Trans_{t}(R) \wedge [Trans_{t}(R)]^2]\!] = [\![[Trans_{t}(R)]^2]\!].$$

Of course, this transition from (2.47) to (2.49) can be iterated using the fact that (2.48) holds true more generally in the form

$$\models [Trans_{t}(R)]^n \rightarrow R^{n+1} \subseteqq_{t} R.$$

Therefore one has for each integer $n > 1$

$$\models Refl(R) \wedge [Trans_{\mathbf{t}}(R)]^n \rightarrow \bigcup_{i=0}^{n+1} R^i \subseteq_{\mathbf{t}} R. \tag{2.50}$$

To finally find the result for $\mathcal{T}_{\mathbf{t}}(R)$ we are looking for, we use the symbolic expression $\prod_{i=0}^{\infty} [Trans_{\mathbf{t}}(R)]^n$ with the truth degree

$$[\![\prod_{i=0}^{\infty} [Trans_{\mathbf{t}}(R)]^n]\!] =_{def} \lim_{n\rightarrow\infty} [\![[Trans_{\mathbf{t}}(R)]^n]\!]$$

and get out of (2.50) according to proposition 2.7 (iv) that

$$\models Refl(R) \wedge \prod_{i=0}^{\infty} [Trans_{\mathbf{t}}(R)]^n \rightarrow \mathcal{T}_{\mathbf{t}}(R) \subseteq_{\mathbf{t}} R$$

and thus

$$\models Refl(R) \wedge \prod_{i=0}^{\infty} [Trans_{\mathbf{t}}(R)]^n \rightarrow \mathcal{T}_{\mathbf{t}}(R) \equiv_{\mathbf{t}} R.$$

Unfortunately, caused by the fact that the expression $\prod_{i=0}^{\infty} [Trans_{\mathbf{t}}(R)]^n$ appears as part of the antecedent, this result seems to be quite week. Only in the special case $\mathbf{t} = \mathbf{t}_G$ does this infinite iteration of the $\mathbf{t}$-conjunction disappear and give

$$\models Refl(R) \wedge Trans_{\mathbf{t}_G}(R) \rightarrow \mathcal{T}_{\mathbf{t}_G}(R) \equiv_{\mathbf{t}_G} R.$$

Chapter 3

Set equations with fuzzy sets

3.1 Fuzzy equations and some of their applications

The notion of the fuzzy equation itself has not so far been defined mathematically. Essentially, in the present literature, three types of fuzzy equations have been discussed. The starting point was in SANCHEZ (1974, 1976) where the author considered fuzzy relation(al) equations of the type

$$R \circ_t X = S \tag{3.1}$$

with $t = t_G = \min$ or, more generally, any intersection operation of a complete Brouwerian lattice and R, S given fuzzy relations. The problem was to find the greatest solution X; the motivation of the problem originated from medical applications of fuzziness; cf. SANCHEZ (1974, 1977) and also KAUFMANN (1977).

A second type of fuzzy equation, also sometimes called fuzzy relation(al) equations and by abuse of notation written $R \circ_t A = B$ or also $A \circ_t R = B$, is equations of the form

$$R''A = B. \tag{3.2}$$

Again, SANCHEZ (1978) was the first one who discussed equations of this type, but in the special form

$$R''A = A$$

which he called "eigen fuzzy sets equations". The fuzzy relation R was supposed to be given, and an eigen fuzzy set A as a solution had to be determined. More precisely, SANCHEZ (1978) looked for the greatest solution of $R''A = A$ because, of course, $A = \emptyset$ is a trivial solution for each R. His motivation, again, came from medical applications.

Fuzzy equations of the kind (3.2) have since found a much wider field of application. In automatic control, for example, the fuzzy controllers first considered by MAMDANI (1974, 1976) turned out to be very useful. In the simplest case such a fuzzy controller has one linguistic input variable and one linguistic output variable. The values of these linguistic variables are fuzzy subsets of corresponding input and output spaces. And the fuzzy controller transforms an input fuzzy set, i.e. some given special value of the input linguistic variable, into an output fuzzy set. Hence, a fuzzy controller may be viewed as a fuzzy relation R connecting input A with output B by a relationship of the form (3.2).

Usually, such a fuzzy controller is supposed to be constructed from a finite list of control rules

$$A_i \models\!\!\Rightarrow B_i, \qquad i = 1, \ldots, n \tag{3.3}$$

connecting values A_i of the input variable with values B_i of the output variable. Hence, the fuzzy controller to be constructed is intended to be described by (and indeed may be identified with) a solution R to the system

$$R''A_i = B_i, \qquad i = 1, \ldots, n \tag{3.4}$$

of fuzzy equations.

Originally, MAMDANI used for R the fuzzy union $\hat{R}$ of the fuzzy cartesian products $A_i \times B_i$. But this fuzzy relation $\hat{R}$ need not be a solution of system (3.4) of fuzzy equations as was mentioned for example by CZOGALA/PEDRYCZ (1981); and conditions which guarantee that this fuzzy relation R is a solution of (3.4) seem to be quite strong, cf. chapter 4. Therefore, the preferable way to construct a fuzzy controller R realizing the system of control rules (3.3) is to choose R as a solution to the system (3.4) of set equations – or at least as some kind of approximate solution to it.

Besides these applications to fuzzy controllers, fuzzy equations of type (3.2) have been discussed in connection with for example the identification of fuzzy systems, prediction in fuzzy systems, and in fuzzy decision-making, cf. for example PEDRYCZ (1983, 1983a).

All these applicational ideas caused a growing interest in the solution of fuzzy equations of type (3.2), in algorithms to construct such solutions, in structural properties of the set of all solutions, and in the solution of systems like (3.4); cf. for example CZOGALA/DREWNIAK/PEDRYCZ (1982), DINOLA/SESSA (1983), PEDRYCZ (1983a), PEDRYCZ/CZOGALA/HIROTA (1984), GOTTWALD (1984b), DINOLA/PEDRYCZ/SESSA (1985), and the book DINOLA/ SESSA/PEDRYCZ/SANCHEZ (1989).

A third type of fuzzy equation is of interest in the field of fuzzy arithmetic, i.e. for the consideration of arithmetical operations for fuzzy numbers. These operations for fuzzy numbers, i.e. for fuzzy subsets of the real line, are defined via the extension principle from the usual arithmetical operations. They are, of course, generalizations of the operations of interval arithmetic, and as in interval arithmetic, the extended subtraction is no longer the inverse operation of extended addition, i.e. for fuzzy numbers A, B the fuzzy equation

$$A +_{t} X = B \tag{3.5}$$

in general does not have the solution $X = B -_{t} A$. Therefore the problem arises to characterize those equations (3.5) which have a solution – e.g. by conditions for A, B – and to determine the set of solutions.

For the existence of direct – additive or multiplicative – inverses this problem was discussed by MIZUMOTO/TANAKA (1979) yielding very strong conditions: only such fuzzy numbers have inverses which are crisp singletons. Therefore, YAGER (1980a) discussed approximate solutions to fuzzy equations and evaluated the quality of approximation with linguistic truth values. Later on SANCHEZ (1984) proved an interesting solvability criterion. And it is this criterion that will be generalized here and also extended to types (3.1), (3.2) of fuzzy equations in such a way that it does not only characterize the fuzzy equations which have solutions – but also in the other cases gives an upper bound measuring the quality of the best possible approximate solutions. Yet, we will measure the quality of an approximate solution not with a linguistic truth value, i.e. a fuzzy subset of [0,1], as was done by YAGER (1980a), but our index of solvability will be a generalized truth value, i.e. a real number of the interval [0,1].

In the following we will refer to fuzzy equations of types (3.1), (3.2) as *fuzzy relational equations* because of the involvement of fuzzy relations, and to fuzzy equations of type

$$A *_{t} X = B$$

as *fuzzy arithmetical equations* because of their (possible) relationship to fuzzy numbers.

3.2 Solvability of fuzzy relational equations

We have first to introduce an additional operation for fuzzy relations that, in a suitable sense, is dual to the relational product $R \circ_t S$ of definition 2.8.

Definition 3.1 *For all fuzzy relations $R, S \in \mathbb{F}(\mathcal{X} \times \mathcal{X})$ and each t-norm t with property $\mathsf{LSC}(t)$ let*

$$R \diamond_t S =_{def} \{(x,y) \,\|\, \forall z((x,z)\,\varepsilon\, R \to_t (z,y)\,\varepsilon\, S)\}.$$

Additionally we need a new operation, besides the fuzzy cartesian product, that gives a fuzzy relation in $\mathbb{F}(\mathcal{X} \times \mathcal{Y})$ out of two fuzzy sets from $\mathbb{F}(\mathcal{X})$ and $\mathbb{F}(\mathcal{Y})$.

Definition 3.2 *For all fuzzy sets $A \in \mathbb{F}(\mathcal{X})$ and $B \in \mathbb{F}(\mathcal{Y})$ and each t-norm t with property $\mathsf{LSC}(t)$ let*

$$A \triangleright_t B =_{def} \{(x,y) \,\|\, x\,\varepsilon\, A \to_t y\,\varepsilon\, B\}.$$

Proposition 3.1 *For all fuzzy sets A, B and fuzzy relations R, S and each t-norm t with property $\mathsf{LSC}(t)$ there hold true*

$$\begin{array}{ll} (i) & \models R''(R \downarrow B) \subseteq_t B, \\ (ii) & \models (A \triangleright_t B)''A \subseteq_t B, \\ (iii) & \models R \circ_t (R \diamond_t S) \subseteq_t S, \\ (iv) & \models (R \diamond_t S^{-1})^{-1} \circ_t R \subseteq_t S. \end{array}$$

Proof. (i) For every point $a \in \mathcal{X}$ of our universe of discourse we have

$$\begin{aligned} [\![a\,\varepsilon\, R''(R \downarrow B)]\!] &= [\![\exists x(x\,\varepsilon\, R \downarrow B \wedge_t (x,a)\,\varepsilon\, R)]\!] \\ &= [\![\exists x((x,a)\,\varepsilon\, R \wedge_t \forall z((x,z)\,\varepsilon\, R \to_t z\,\varepsilon\, B))]\!] \\ &\leq [\![\exists x((x,a)\,\varepsilon\, R \wedge_t ((x,a)\,\varepsilon\, R \to_t a\,\varepsilon\, B))]\!] \\ &\leq [\![\exists x(a\,\varepsilon\, B)]\!] = [\![a\,\varepsilon\, B]\!] \end{aligned}$$

by definition (1.1) of many-valued generalization, the monotonicity (T2) of every t-norm, and proposition 1.17 (i). Our result follows by the rule of generalization and proposition 1.13.

All the other facts will be proved along the same lines of argument: to get the analogous inequalities use corresponding definitions, delete many-valued generalization, and reduce à la proposition 1.17 (i). Thus we can omit the details. QED

Proposition 3.2 *For all fuzzy sets A, B and fuzzy relations R, S, T and each t-norm t with property $\mathsf{LSC}(t)$ there hold true*

$$
\begin{array}{ll}
(i) & \models R''A \subseteqq_t B \rightarrow_t A \subseteqq_t R \downarrow B, \\
(ii) & \models R''A \subseteqq_t B \rightarrow_t R \subseteqq_t A \rhd_t B, \\
(iii) & \models R \circ_t S \subseteqq_t T \rightarrow_t S \subseteqq_t R^{-1} \diamond_t T, \\
(iv) & \models R \circ_t S \subseteqq_t T \rightarrow_t R \subseteqq_t (S \diamond_t T^{-1})^{-1}.
\end{array}
$$

Proof. (i) The following estimation proves the result:

$$
\begin{array}{rcl}
[\![A \subseteqq_t R \downarrow B]\!] & = & [\![\forall x(x \varepsilon A \rightarrow_t x \varepsilon R \downarrow B)]\!] \\
& = & [\![\forall x \forall y(x \varepsilon A \rightarrow_t ((x,y) \varepsilon R \rightarrow_t y \varepsilon B))]\!] \\
& \geq & [\![\forall x \forall y(x \varepsilon A \wedge_t (x,y) \varepsilon R \rightarrow_t y \varepsilon B)]\!] \\
& \geq & [\![\forall y(\exists x(x \varepsilon A \wedge_t (x,y) \varepsilon R) \rightarrow_t y \varepsilon B)]\!] \\
& = & [\![\forall y(y \varepsilon R''A \rightarrow_t y \varepsilon B)]\!] \\
& = & [\![R''A \subseteqq_t B]\!].
\end{array}
$$

Here, besides the corresponding definitions, the propositions 1.27 (i), 1.19 (i), and 1.27 (ii) have been successively used .

Just the same line of argument is also successful for parts (ii) to (iv). Thus again the details can be omitted. QED

Corollary 3.3 *For all fuzzy sets A, B and fuzzy relations R, S, T and each t-norm t with the property $\mathsf{LSC}(t)$ the following facts hold true:*

(*i*) *Every solution of equation $R''X \equiv_t B$ is a fuzzy subset of $R \downarrow B$.*

(*ii*) *Every solution of equation $X''A \equiv_t B$ is a fuzzy subset of $A \diamond_t B$.*

(*iii*) *Every solution of equation $R \circ_t X \equiv_t S$ is a fuzzy subset of $R^{-1} \diamond_t S$.*

(*iv*) *Every solution of equation $X \circ_t R \equiv_t S$ is a fuzzy subset of $(R \diamond_t S^{-1})^{-1}$.*

Proof. Suppose A to be a solution of the equation $R''X \equiv_t B$. Then of course $[\![R''A \equiv_t B]\!] = 1$ and therefore also $[\![R''A \subseteqq_t B]\!] = 1$, i.e. by proposition 3.2 (i): $[\![A \subseteqq_t R \downarrow B]\!] = 1$. Thus (i) is proved.

The other claims follow in the same manner. QED

As a side remark we have to note that we have misused the term "fuzzy subset" here a little, otherwise in this book a fuzzy subset of some (crisp!) set $\mathcal{X}$ will be taken to mean any function from $\mathcal{X}$ into [0,1]. But here by a fuzzy subset A of a fuzzy (!) set B there is meant such a fuzzy set A for which $\models A \subseteqq_t B$ holds true. This difference will always become clear from the context, and we hope not to cause any misunderstanding.

Now we are able to state and prove the main results for (single) fuzzy relational equations.

Theorem 3.4 *Suppose that* $\mathbf{t}$ *is a t-norm with property* $\mathsf{LSC}(\mathbf{t})$. *Then for all fuzzy sets* A, B *and fuzzy relations* R, S *there hold true*

$$
\begin{array}{ll}
(i) & \models \exists X(R''X \equiv_t B) \leftrightarrow_t R''(R \downarrow B) \equiv_t B, \\
(ii) & \models \exists X(X''A \equiv_t B) \leftrightarrow_t (A \rhd_t B)''A \equiv_t B, \\
(iii) & \models \exists X(R \circ_t X \equiv_t S) \leftrightarrow_t R \circ_t (R^{-1} \diamond_t S) \equiv_t S, \\
(iv) & \models \exists X(X \circ_t R \equiv_t S) \leftrightarrow_t (R \diamond_t S^{-1})^{-1} \circ_t R \equiv_t S.
\end{array}
$$

Proof. (i) By proposition 1.13 we have to prove the equality

$$[\![\exists X(R''X \equiv_t B)]\!] = [\![R''(R \downarrow B) \equiv_t B]\!]$$

for these truth degrees. Obviously, from the definition (1.2) of the existential quantifier and the fact that $R \downarrow B \in \mathbb{F}(\mathcal{X})$, we immediately have

$$[\![\exists X(R''X \equiv_t B)]\!] \geq [\![R''(R \downarrow B) \equiv_t B]\!]$$

Thus the reverse inequality is what really has to be proved. But, from proposition 3.1 (i) and definition 2.3 one has

$$[\![R''(R \downarrow B) \equiv_t B]\!] = [\![B \subseteqq_t R''(R \downarrow B)]\!]$$

and hence only to prove

$$[\![\exists X(R''X \equiv_t B)]\!] \geq [\![B \subseteqq_t R''(R \downarrow B)]\!].$$

For this it is enough to show for each fuzzy set A

$$[\![R''A \equiv_t B]\!] \geq [\![B \subseteqq_t R''(R \downarrow B)]\!].$$

Using definition 2.3 and proposition 3.2 (ii) one has

$$
\begin{aligned}
[\![R''A \equiv_t B]\!] &= [\![R''A \subseteqq_t B \wedge_t B \subseteqq_t R''A]\!] \\
&\leq [\![A \subseteqq_t R \downarrow B \wedge_t B \subseteqq_t R''A]\!]
\end{aligned}
$$

and furthermore by the monotonicity of the full image and the transitivity of the fuzzy inclusion $\subseteq_t$, i.e. by propositions 2.18 (i) and 2.2 (iii) one is able to continue this estimation as

$$\begin{aligned}[\![R''A \equiv_t B]\!] &\leq [\![R''A \subseteq_t R''(R \downarrow B) \wedge_t B \subseteq_t R''A]\!] \\ &\leq [\![B \subseteq_t R''(R \downarrow B)]\!].\end{aligned}$$

Thus (i) is proved.

The other results formulated in this theorem can be proved in the same way. All the facts necessary for carrying out these proofs have been formulated in propositions 2.2, 2.18, 2.15, 3.1, and 3.2. Hence, the rest of this proof is now again routine. QED

Corollary 3.5 *For all fuzzy sets A, B and fuzzy relations R, S, T and each t-norm t with the property $\mathsf{LSC}(t)$ the following facts hold true:*

(*i*) *Equation $R''X = B$ has a solution iff $R''(R \downarrow B) = B$.*

(*ii*) *Equation $X''A = B$ has a solution iff $(A \diamond_t B)''A = B$.*

(*iii*) *Equation $R \circ_t X = S$ has a solution iff $R \circ_t (R^{-1} \diamond_t S) = S$.*

(*iv*) *Equation $X \circ_t R = S$ has a solution iff $(R \diamond_t S^{-1})^{-1} \circ_t R = S$.*

Proof. (i) If $R''(R \downarrow B) = B$ then $R \downarrow B$ is a solution to equation $R''X = B$; and then it is the greatest one by corollary 3.3 (i). If otherwise equation $R''X = B$ has a solution, say A, then $R''A = B$, i.e. $[\![R''A \equiv_t B]\!] = 1$ and hence $[\![\exists X(R''X \equiv_t B)]\!] = 1$, thus $R''(R \downarrow B) \equiv_t B$ by proposition 3.4 (i) and this means $R''(R \downarrow B) = B$ because of (2.14).

(ii) to (iv) follow by corresponding arguments. QED

3.3 Solvability of fuzzy arithmetical equations

Now we suppose that in the universe of discourse $\mathcal{X}$ a binary operation $*$ is given. This operation is extended to a binary operation, denoted $*_t$ for a given t-norm t, in the class of all fuzzy subsets of $\mathcal{X}$ by the extension principle. Additionally, a companion operation $\tilde{*}_t$ is needed which in the present case is analogous to the earlier operations $\triangleright_t$, $\diamond_t$, as well as $\downarrow$ introduced in definitions 3.2, 3.1, and 2.9. (For the special case that $t = t_G = \min$ this operation was introduced by SANCHEZ (1984).)

Definition 3.3 *For all fuzzy sets A, B and each t-norm t with the property* $\mathsf{LSC}(t)$ *put*

$$B \,\tilde{*}_t\, A =_{def} \{y \,\|\, \forall x(x \,\varepsilon\, A \rightarrow_t x * y \,\varepsilon\, B)\}.$$

Proposition 3.6 *For all fuzzy sets A, B, C and each t-norm t with the property* $\mathsf{LSC}(t)$ *there hold true*

$$\begin{array}{ll} (i) & \models B \subseteq_t C \rightarrow_t A *_t B \subseteq_t A *_t C, \\ (ii) & \models A *_t B \subseteq_t C \rightarrow_t B \subseteq_t C \,\tilde{*}_t\, A, \\ (iii) & \models A *_t (B \,\tilde{*}_t\, A) \subseteq_t B. \end{array}$$

Proof. (i) To get the result we prove for the corresponding truth degrees the inequality

$$[\![B \subseteq_t C]\!] \leq [\![A *_t B \subseteq_t A *_t C]\!].$$

Immediately from the extension principle together with (2.24) one has

$$\begin{aligned} [\![A *_t B \subseteq_t A *_t C]\!] &= [\![\forall x(x \,\varepsilon\, A *_t B \rightarrow_t x \,\varepsilon\, A *_t C)]\!] \\ &= [\![\forall x(\exists y \exists z(y \,\varepsilon\, A \wedge_t z \,\varepsilon\, B \wedge_t x \doteq y * z) \rightarrow_t x \,\varepsilon\, A *_t C)]\!] \\ &= [\![\forall x \forall y \forall z(y \,\varepsilon\, A \wedge_t z \,\varepsilon\, B \wedge_t x \doteq y * z \rightarrow_t x \,\varepsilon\, A *_t C)]\!]. \end{aligned}$$

With the same trick as in the proof of proposition 2.18 (i) for eliminating the existential quantifier that comes into this formula by substituting $x \,\varepsilon\, A *_t C$ through $\exists u \exists v(u \,\varepsilon\, A \wedge_t v \,\varepsilon\, C \wedge_t x \doteq u * v)$ we get furthermore

$$\begin{aligned} &[\![A *_t B \subseteq_t A *_t C]\!] \\ &\qquad \geq [\![\forall x \forall y \forall z(z \,\varepsilon\, B \wedge_t \Theta(x,y,z) \rightarrow_t z \,\varepsilon\, C \wedge_t \Theta(x,y,z)]\!] \end{aligned}$$

with formula $\Theta(x, y, z) : y \,\varepsilon\, A \wedge_t x \doteq y * z$. Now proposition 1.18 (i) and the dropping of the empty quantifiers allows us to continue the estimation

$$[\![A *_t B \subseteq_t A *_t C]\!] \geq [\![\forall z(z \,\varepsilon\, B \rightarrow_t z \,\varepsilon\, C]\!] = [\![B \subseteq_t C]\!]$$

and thus to finish the proof of (i).

(ii) The corresponding inequality for truth values will now also be proved. It is

$$\begin{aligned} [\![B \subseteq_t C \,\tilde{*}_t\, A]\!] &= [\![\forall x(x \,\varepsilon\, B \rightarrow_t \forall y(y \,\varepsilon\, A \rightarrow_t y * x \,\varepsilon\, C))]\!] \\ &= [\![\forall x \forall y(x \,\varepsilon\, B \wedge_t y \,\varepsilon\, A \rightarrow_t y * x \,\varepsilon\, C)]\!] \end{aligned}$$

which because of $[\![x \varepsilon B \wedge_t y \varepsilon A]\!] \leq [\![y * x \varepsilon A *_t B]\!]$ and the antimonotonicity of φ_t in the first argument can be continued as

$$\begin{aligned}[\![B \subseteqq_t C \tilde{*}_t A]\!] &\geq [\![\forall x \forall y(y * x \varepsilon A *_t B \rightarrow_t y * x \varepsilon C)]\!] \\ &\geq [\![\forall z(z \varepsilon A *_t B \rightarrow_t z \varepsilon C)]\!] \\ &= [\![A *_t B \subseteqq_t C]\!].\end{aligned}$$

Finally, (iii) is proved by simply applying the corresponding definitions together with proposition 1.17 (i). QED

Theorem 3.7 *For each t-norm t with the property* LSC(t) *and all fuzzy sets* A, B *there holds true*

$$\models \exists X(A *_t X \equiv_t B) \leftrightarrow_t A *_t (B \tilde{*}_t A) \equiv_t B.$$

Proof. Obviously

$$[\![A *_t (B \tilde{*}_t A) \equiv_t B]\!] \leq [\![\exists X(A *_t X \equiv_t B)]\!]$$

and by proposition 3.6 (iii) also

$$[\![A *_t (B \tilde{*}_t A) \equiv_t B]\!] = [\![B \subseteqq_t A *_t (B \tilde{*}_t A)]\!].$$

Therefore, it remains to be proved

$$[\![\exists X(A *_t X \equiv_t B)]\!] \leq [\![B \subseteqq_t A *_t (B \tilde{*}_t A)]\!],$$

i.e. for every fuzzy set C that

$$[\![A *_t C \equiv_t B]\!] \leq [\![B \subseteqq_t A *_t (B \tilde{*}_t A)]\!].$$

And indeed

$$\begin{aligned}[\![A *_t C \equiv_t B]\!] &= [\![A *_t C \subseteqq_t B \wedge_t B \subseteqq_t A *_t C]\!] \\ &\leq [\![C \subseteqq_t B \tilde{*}_t A \wedge_t B \subseteqq_t A *_t C]\!] \\ &\leq [\![A *_t C \subseteqq_t A *_t (B \tilde{*}_t A) \wedge_t B \subseteqq_t A *_t C]\!] \\ &\leq [\![B \subseteqq_t A *_t (B \tilde{*}_t A)]\!]\end{aligned}$$

by the previous results. QED

For completeness we should, for non-commutative operations $*$, also consider equations of the type

$$X *_t A \equiv_t B$$

but by considering in $\mathcal{X}$ the operation $a \sharp b = b * a$ we may reduce this case to that which is considered in theorem 3.7.

Proposition 3.8 *Suppose that the t-norm t has the property* $\mathsf{LSC}(t)$. *Then equation $A *_t X = B$ has a solution iff $A *_t (B \tilde{*}_t A) = B$; and if equation $A *_t X = B$ has a solution, then $B \tilde{*}_t A$ is the greatest one.*

Proof. By the same argument that was used in the proofs of corollaries 3.5 and 3.3. QED

The result of this corollary 3.8 was, for the t-norm $t = \min$, first proved by SANCHEZ (1984) and the starting point for the present generalizations.

3.4 Solvability of systems of fuzzy equations

So far we have only discussed the solvability of single fuzzy equations, despite the fact that systems of fuzzy relational equations like (3.4) are essential means for the construction of fuzzy controllers. Moreover, the research literature till now has not paid much attention to this topic – GOTTWALD (1984b) seems to be an exception. Here there are given for systems of fuzzy equations, of each one of the types considered before, inequalities for the truth degrees of the formula of our language of many-valued logic that states the existence of a solution to the system.

Unfortunately, we have not found equations characterizing these truth degrees as in theorems 3.4 and 3.7, but only inequalities. Nevertheless we find lower bounds – and in a certain weak sense also upper bounds.

Theorem 3.9 *Suppose that the t-norm t has the property* $\mathsf{LSC}(t)$. *Consider finite sequences $(A_i)_{1 \leq i \leq n}$ and $(B_i)_{1 \leq i \leq n}$ of fuzzy sets as well as $(R_i)_{1 \leq i \leq n}$ and $(S_i)_{1 \leq i \leq n}$ of fuzzy relations. Then there hold true*

(*i*) *in the case of a system of fuzzy equations with given fuzzy relations and given full images that has to be solved for an argument fuzzy set:*

$$
\begin{aligned}
[\![(\exists X \prod_{i=1}^{n}(R_i''X \equiv_t B_i))^n]\!]
&\leq [\![\prod_{i=1}^{n}(R_i''(\bigcap_{j=1}^{n}(R_j \downarrow B_j)) \equiv_t B_i)]\!] \\
&\leq [\![\exists X \prod_{i=1}^{n}(R_i''X \equiv_t B_i)]\!]
\end{aligned}
$$

(ii) *in the case of a system of fuzzy equations with given input fuzzy sets and given full images that has to be solved for a corresponding fuzzy relation:*

$$\begin{aligned}
&[\![(\exists X \prod_{i=1}^{n}(X''A_i \equiv_t B_i))^n]\!] \\
&\quad \leq [\![\prod_{i=1}^{n}((\bigcap_{j=1}^{n}(A_j \rhd_t B_j))''A_i \equiv_t B_i)]\!] \\
&\quad\quad \leq [\![\exists X \prod_{i=1}^{n}(X''A_i \equiv_t B_i)]\!]
\end{aligned}$$

(iii) *in the case of a system of fuzzy equations with given relational products and given first factors that has to be solved for the second factor:*

$$\begin{aligned}
&[\![(\exists X \prod_{i=1}^{n}(R_i \circ_t X \equiv_t S_i))^n]\!] \\
&\quad \leq [\![\prod_{i=1}^{n}(R_i \circ_t \bigcap_{j=1}^{n}(R_j^{-1} \diamond_t S_j) \equiv_t S_i)]\!] \\
&\quad\quad \leq [\![\exists X \prod_{i=1}^{n}(R_i \circ_t X \equiv_t S_i)]\!]
\end{aligned}$$

(iv) *in the case of a system of fuzzy equations with given relational products and given second factors that has to be solved for the first factor:*

$$\begin{aligned}
&[\![(\exists X \prod_{i=1}^{n}(X \circ_t R_i \equiv_t S_i))^n]\!] \\
&\quad \leq [\![\prod_{i=1}^{n}((\bigcap_{j=1}^{n}(R_j \diamond_t S_j^{-1})^{-1}) \circ_t R_i \equiv_t S_i)]\!] \\
&\quad\quad \leq [\![\exists X \prod_{i=1}^{n}(X \circ_t R_i \equiv_t S_i)]\!]
\end{aligned}$$

(v) *in the case of a system of "linear" fuzzy arithmetical equations with given first operands and given results that has to be solved for the second operand:*

$$[\![(\exists X \prod_{i=1}^{n}(A_i *_t X \equiv_t B_i))^n]\!]$$

$$\leq [\prod_{i=1}^{n}(A_i *_t (\bigcap_{j=1}^{n}(B_j \tilde{*}_t A_j)) \equiv_t B_i)]$$

$$\leq [\exists X \prod_{i=1}^{n}(A_i *_t X \equiv_t B_i)]$$

Proof. As before the proofs run parallel for each one of these cases. We therefore give only the derivation for case (ii) because of its close connection with fuzzy controllers.

Because of proposition 1.31 one has

$$[(\exists X \prod_{i=1}^{n}(X''A_i \equiv_t B_i))^n]$$

$$= \sup\{ [(\prod_{i=1}^{n}(T''A_i \equiv_t B_i))^n] \mid T \in I\!F(\mathcal{X} \times \mathcal{X})\},$$

hence for the first inequality stated in (ii) it has only to be proved

$$[(\prod_{i=1}^{n}(T''A_i \equiv_t B_i))^n] \leq [\prod_{i=1}^{n}((\bigcap_{j=1}^{n}(A_j \rhd_t B_j))''A_i \equiv_t B_i)] \tag{3.6}$$

for any $T \in I\!F(\mathcal{X} \times \mathcal{X})$. For the sake of brevity let be

$$D = \bigcap_{j=1}^{n}(A_j \rhd_t B_j).$$

This gives the following sequence of inequalities:

$$[(\prod_{i=1}^{n}(T''A_i \equiv_t B_i))^n]$$

$$= [\prod_{i,j=1}^{n}(T''A_i \subseteq_t B_i \wedge_t B_i \subseteq_t T''A_i)]$$

$$= [\prod_{i,j=1}^{n}(T''A_i \subseteq_t B_i) \wedge_t \prod_{i,j=1}^{n}(B_i \subseteq_t T''A_i)]$$

$$\leq [\prod_{i,j=1}^{n}(T \subseteq_t A_i \rhd_t B_i) \wedge_t \prod_{i,j=1}^{n}(B_i \subseteq_t T''A_i)]$$

$$\leq [\prod_{j=1}^{n}(\bigwedge_{i=1}^{n}(T \subseteq_t A_i \rhd_t B_i)) \wedge_t \prod_{i,j=1}^{n}(B_i \subseteq_t T''A_i)]$$

because of proposition 3.2 (ii) and the fact that $x\mathbf{t}y \leq \min(x, y)$ for every t-norm $\mathbf{t}$ and all $x, y \in [0,1]$, which because of proposition 2.7 (iii) gives furthermore

$$\leq [\![\prod_{j=1}^{n}(T \subseteqq_{\mathbf{t}} D) \wedge_{\mathbf{t}} \prod_{i,j=1}^{n}(B_i \subseteqq_{\mathbf{t}} T''A_i)]\!]$$

$$\leq [\![\prod_{i=1}^{n}(B_i \subseteqq_{\mathbf{t}} T''A_i \wedge_{\mathbf{t}} T \subseteqq_{\mathbf{t}} tD)]\!]$$

$$\leq [\![\prod_{i=1}^{n}(B_i \subseteqq_{\mathbf{t}} T''A_i \wedge_{\mathbf{t}} T''A_i \subseteqq_{\mathbf{t}} tD''A_i)]\!]$$

$$\leq [\![\prod_{i=1}^{n}(B_i \subseteqq_{\mathbf{t}} D''A_i)]\!],$$

where now successively $x\mathbf{t}y \leq x$ for each t-norm $\mathbf{t}$ and all $x, y \in [0,1]$, proposition 2.18 (ii), and the generalized transitivity of the fuzzy inclusion $\subseteqq_{\mathbf{t}}$, i.e. proposition 2.2 (iii) have been used. From proposition 2.7 (i) we get for every $1 \leq i \leq n$

$$\models D \subseteqq_{\mathbf{t}} A_i \rhd_{\mathbf{t}} B_i$$

and therefore by propositions 2.18 (ii), 3.1 (ii), and again the generalized transitivity of the fuzzy inclusion $\subseteqq_{\mathbf{t}}$ immediately

$$\models D''A_i \subseteqq_{\mathbf{t}} B_i.$$

This gives

$$[\![\prod_{i=1}^{n}(D''A_i \subseteqq_{\mathbf{t}} B_i)]\!] = 1$$

and hence

$$[\![\prod_{i=1}^{n}(B_i \subseteqq_{\mathbf{t}} D''A_i)]\!] = [\![\prod_{i=1}^{n}(D''A_i \equiv_{\mathbf{t}} B_i)]\!].$$

So, finally, inequality (3.6) is proved. And for the full proof of (ii) only

$$[\![\prod_{i=1}^{n}(D''A_i \equiv_{\mathbf{t}} B_i)]\!] \leq [\![\exists X \prod_{i=1}^{n}(X''A_i \equiv_{\mathbf{t}} B_i)]\!]$$

remains. But this is obvious. Thus, the proof of (ii) is complete. QED

Because of their relationship with fuzzy controllers and with fuzzy arithmetic, in the following corollaries we discuss only systems of fuzzy equations of the type

$$X''A_i \equiv_t B_i, \qquad 1 \leq i \leq n \tag{3.7}$$

and of the type

$$A_i * X \equiv_t B_i, \qquad 1 \leq i \leq n. \tag{3.8}$$

Corollary 3.10 *Consider only the t-norm $t = t_G = \min$ and omit all indices that refer to this t-norm. Then we have for the systems (3.7), (3.8) of fuzzy equations the solvability criteria:*

(i) in the case of a system of fuzzy equations with given input fuzzy sets and given full images that has to be solved for a corresponding fuzzy relation:

$$\models \exists X \bigwedge_{i=1}^{n} (X''A_i \equiv_t B_i) \leftrightarrow \bigwedge_{i=1}^{n} ((\bigcap_{j=1}^{n} (A_j \rhd_t B_j))''A_i \equiv_t B_i);$$

(ii) in the case of a system of "linear" fuzzy arithmetical equations with given first operands and given results that has to be solved for the second operand:

$$\models \exists X \bigwedge_{i=1}^{n} (A_i *_t X \equiv_t B_i) \leftrightarrow \bigwedge_{i=1}^{n} (A_i *_t (\bigcap_{j=1}^{n} (B_j \tilde{*}_t A_j)) \equiv_t B_i).$$

Proof. For the t-norm $t = \min$ we obviously have $[\![\Theta^n]\!] = [\![\Theta]\!]$ for each well-formed formula Θ, hence (i) and (ii) are only reformulations of theorem 3.9 (ii) and 3.9 (v) respectively for the present case. QED

Corollary 3.11 *Suppose that the t-norm t has property $\mathsf{LSC}(t)$. Then there hold true:*

(i) System (3.7) has a solution iff $\bigcap_{i=1}^{n}(A_i \rhd_t B_i)$ is a solution to this system; and if (3.7) has a solution then $\bigcap_{i=1}^{n}(A_i \rhd_t B_i)$ is the greatest one.

(ii) System (3.8) has a solution iff $\bigcap_{i=1}^{n}(B_i \tilde{*}_t A_i)$ is a solution to this system; and if (3.8) has a solution then $\bigcap_{i=1}^{n}(B_i \tilde{*}_t A_i)$ is the greatest one.

Proof. We prove only (i) because the proof of (ii) is similar. Put $D = \bigcap_{i=1}^{n}(A_i \rhd_t B_i)$. If D is a solution of (3.7), then (3.7) obviously has a solution. Hence suppose that (3.7) has a solution, say R. Then for each $1 \leq i \leq N$

$$[\![R''A_i \equiv_t B_i]\!] = 1$$

and thus by proposition 3.2 (ii)

$$[\![\bigwedge_{i=1}^{n}(R \subseteqq_t A_i \rhd_t B_i)]\!] = 1$$

which by proposition 2.7 (iii) gives $[\![R \subseteqq_t D]\!] = 1$, i.e. the fuzzy set R is a (fuzzy) subset of the fuzzy set D. But also

$$[\![\prod_{i=1}^{n}(R''A_i \equiv_t B_i)]\!] = 1$$

and therefore

$$[\![(\exists X \prod_{i=1}^{n}(X''A_i \equiv_t B_i))^n]\!] = [\![\exists X \prod_{i=1}^{n}(X''A_i \equiv_t B_i)]\!] = 1.$$

Now by theorem 3.9 (ii) one has

$$[\![\prod_{i=1}^{n}(D''A_i \equiv_t B_i)]\!] = 1$$

which means that D is a solution to (3.7). Hence (i) is proved. QED

3.5 Solvability degrees and approximate solutions

In a general setting, in referring to a fuzzy (relation) equation one quite often has in mind an equation describing a relationship between fuzzy sets in two (possibly different) "spaces", i.e. universes of discourse. Such a form of relationship is supposed to be represented by a fuzzy relation between the elements of these "spaces", i.e. over the (crisp) cartesian product of those universes of discourse. More precisely, consider two fuzzy sets $A \in \mathbb{F}(\mathcal{X})$ and $B \in \mathbb{F}(\mathcal{Y})$, as well as a fuzzy relation $R \in \mathbb{F}(\mathcal{X} \times \mathcal{Y})$. Then a fuzzy relation equation can be written down in a general form as

$$\Theta(R, A) = B \tag{3.9}$$

where Θ is a suitable operator producing a fuzzy set B out of a fuzzy set A and a fuzzy relation R. With respect to our previous discussion the actual situation is a quite direct generalization of equations of type (3.2). And the case of a system

$$\Theta(R, A_i) = B_i, \qquad i = 1, \ldots, n \tag{3.10}$$

of fuzzy relation equations fits into these considerations as well.

Even more general, of course, is to consider Θ as an operator (of some finite arity) which maps fuzzy sets and relations onto fuzzy sets or fuzzy relations – and for which some of the arguments have to be determined. We will not discuss the problem of fuzzy equations in this generality.

The following considerations are mainly restricted to the case of fuzzy relation equations of type (3.2) and of systems (3.4) of such equations, i.e. to special cases of (3.9) and (3.10). They can be and sometimes will be extended to other types of fuzzy relation equations, as well as to fuzzy arithmetical equations without principal difficulties. The last section has shown how far-reaching the analogies are which one has in treating these "parallel" cases guided by mainly the same ideas. Hence we essentially avoid that kind of "parallelization" here.

Our strategy in the present chapter toward discussing the solvability behaviour of fuzzy (relation) equations is essentially that one of a "many-valued translation" — we changed, with respect to some t-norm $\boldsymbol{t}$, from traditional equations like (3.9): $B = \Theta(R, A)$ which are to be solved for R or A to their many-valued counterparts $B \equiv_{\boldsymbol{t}} \Theta(R, A)$ which – for lower semicontinuous t-norms $\boldsymbol{t}$ – have the property that always

$$[\![\Theta(R, A) \equiv_{\boldsymbol{t}} B]\!] = 1 \Leftrightarrow \Theta(R, A) = B.$$

What we achieved to different levels of satisfaction were characterizations of the truth degrees $[\![\exists X(\Theta(R, X) \equiv_{\boldsymbol{t}} B)]\!]$ and $[\![\exists X(\Theta(X, A) \equiv_{\boldsymbol{t}} B)]\!]$ which did not involve the variable[1] X, i.e. which are built up using only the "given data" R, B and A, B respectively.

For systems of fuzzy equations the situation, in the last section, was almost the same: only the truth degree to be determined now was for example

[1]In some sense that variable changes its character in those formulas: in the first one it is a variable for a fuzzy subset of an universe of discourse $\mathcal{X}$ and in the second one for a fuzzy relation, that means for a fuzzy subset of an universe $\mathcal{X} \times \mathcal{Y}$. But this distinction is clear from the context and thus there is no need to express it in the notation too.

of the form $[\![\exists X \prod_{i=1}^{n} (\Theta(X, A_i) \equiv_t B_i)]\!]$, i.e. it was taken as the truth degree of the sentence[2]: *"The system of fuzzy equations*

$$\Theta(X, A_i) \equiv_t B_i, \qquad i = 1, \ldots, n$$

has a solution."

Instead of directly discussing the problem of solvability of fuzzy relation equations (3.9) or of systems (3.10) of such equations we consider the truth degrees which we just mentioned as *solvability degrees* indicating the solvability behaviour of our (systems of) fuzzy equations.

Definition 3.4 *For each one of the fuzzy (relation) equations (3.9) and of the systems (3.10) of such equations, which are supposed to be solved with respect to the fuzzy relation R, their* solvability degree *is the truth degree*

$$\xi_0 =_{def} [\![\exists X(\Theta(X, A) \equiv_t B)]\!] \tag{3.11}$$

and in case of a system of equations

$$\xi =_{def} [\![\exists X \prod_{i=1}^{n} (\Theta(X, A_i) \equiv_t B_i)]\!]. \tag{3.12}$$

As an immediate consequence we have the following corollary. To formulate it we introduce in analogy with definition 1.9 the finite iteration $\mathbf{T}$ of the t-norm t in the following way for any $u_1, \ldots, u_{n+1} \in [0, 1]$:

$$\mathop{\mathbf{T}}_{i=1}^{1} u_i =_{def} u_1, \qquad \mathop{\mathbf{T}}_{i=1}^{n+1} u_i =_{def} (\mathop{\mathbf{T}}_{i=1}^{n} u_i) \, t \, u_{n+1}. \tag{3.13}$$

Corollary 3.12 *For all $A, A_i \in I\!F(\mathcal{X})$ and $B, B_i \in I\!F(\mathcal{Y})$ the solvability degree ξ_0 of equation (3.9) and the solvability degree ξ of the system (3.10) of equations are*

$$\xi_0 = \sup\{ [\![\Theta(R, A) \equiv_t B]\!] \mid R \in I\!F(\mathcal{X} \times \mathcal{Y})\}$$

$$\xi = \sup\{ \mathop{\mathbf{T}}_{i=1}^{n} [\![\Theta(R, A_i) \equiv_t B_i]\!] \mid R \in I\!F(\mathcal{X} \times \mathcal{Y})\}.$$

[2]Of course, this sentence has to be read as being formulated in the language of many-valued logic.

What, now, is the relation between the solvability of equations and the value of the solvability degree? In one direction there is quite a simple connection.

Proposition 3.13 *If a fuzzy equation (3.9) or a system (3.10) of such equations has a solution, then its solvability degree is* $= 1$.

Proof. Obvious from proposition 2.4 (*i*).

Thus, a solvability degree < 1 indicates that there is no solution.

For the converse case that the solvability degree is $=1$ only a partial solution is at hand.

Suppose that a t-norm $\boldsymbol{t}$ is given such that there is a finite set $L \subseteq [0,1]$ of truth degrees with $0, 1 \in L$, and which is closed with respect to the operations min, max and $u \mapsto 1-u$, as well as with respect to $\boldsymbol{t}$. Then $\boldsymbol{t}$ can be supposed to be a *continuous* t-norm and L is even closed with respect to the operation $\varphi_{\boldsymbol{t}}$. Let us call such a finite set L a (*finite*) ***t**-clan*. By $\mathbb{F}_L(\mathcal{X})$ we denote the class of all L-fuzzy subsets of $\mathcal{X}$.

Given a continuous t-norm $\boldsymbol{t}$ and a finite $\boldsymbol{t}$-clan L, for all $A, A_i \in \mathbb{F}_L(\mathcal{X})$ and $B, B_i \in \mathbb{F}_L(\mathcal{Y})$ we also consider the *relative* solvability degrees

$$\xi_0^{(L)} =_{def} \sup\{ [\![\Theta(R,A) \equiv_{\boldsymbol{t}} B]\!] \mid R \in \mathbb{F}_L(\mathcal{X} \times \mathcal{Y})\} \tag{3.14}$$

$$\xi^{(L)} =_{def} \sup\{ \mathop{\mathbf{T}}_{i=1}^{n} [\![\Theta(R,A_i) \equiv_{\boldsymbol{t}} B_i]\!] \mid R \in \mathbb{F}_L(\mathcal{X} \times \mathcal{Y})\}. \tag{3.15}$$

Of course, using bounded quantification and writing $\mathcal{R} = \mathbb{F}_L(\mathcal{X} \times \mathcal{Y})$ one has

$$\xi_0^{(L)} = [\![\exists_{\mathcal{R}} X(\Theta(X,A) \equiv_{\boldsymbol{t}} B)]\!]$$

$$\xi^{(L)} = [\![\exists_{\mathcal{R}} X \prod_{i=1}^{n} (\Theta(X,A_i) \equiv_{\boldsymbol{t}} B_i)]\!].$$

Proposition 3.14 *Suppose that L is a finite **t**-clan with respect to a continuous t-norm **t**. Then one has*

$$\xi_0^{(L)} \leq \xi_0 \qquad \textit{and} \qquad \xi^{(L)} \leq \xi$$

for all such (systems of) fuzzy equations with $A, A_i \in \mathbb{F}_L(\mathcal{X})$ and $B, B_i \in \mathbb{F}_L(\mathcal{Y})$.

Proof. The "full" solvability degrees ξ_0, ξ are, according to corollary 3.12, suprema over bigger classes of fuzzy relations than the relative solvability degrees, and hence not smaller numbers than those degrees. QED

Proposition 3.15 *Suppose that L is a finite **t**-clan with respect to a continuous t-norm **t** and that $\xi_0^{(L)} = 1$ resp. $\xi^{(L)} = 1$ for some (system of) fuzzy equation(s). Then that fuzzy equation or that system of fuzzy equations has a solution.*

Proof. Because of the finiteness of the t-clan L there exists an L-fuzzy relation R_0 such that $[\![\Theta(R_0, A) \equiv_t B]\!] = 1$ resp. $\mathbf{T}_{i=1}^{n} [\![\Theta(R_0, A_i) \equiv_t B_i]\!] = 1$. In the case of a single equation that means $\Theta(R, A) = B$ by (2.14), and for a system of fuzzy equations that first means $[\![\Theta(R, A_i) \equiv_t B_i]\!] = 1$ for each $i = 1, \ldots, n$ by (T2), (T3), and thus also that always $\Theta(R, A_i) = B_i$. QED

From a theoretical point of view it is unsatisfactory to have these restrictions to the case of finite t-clans. Indeed, the guess is that the last proposition holds true without this assumption. Yet, in practice this restriction to finite t-clans is not such a severe one because quite often one considers only finite universes of discourse and also t-norms for which a suitable subset L_m of $[0,1]$ of the kind $L_m = \{\frac{k}{m-1} \mid 0 \leq k < m\}$, i.e. of equidistant rational numbers is a t-clan.

The interest in the solvability degree of (systems of) fuzzy equations does not only come from this connection between the existence of solutions and the value of this degree being =1. There is another aspect that deserves to be taken into consideration: a connection between the solvability index and the approximate solvability of (systems of) fuzzy equations.

The background for this new aspect is that the generalized, "fuzzified" identity relations $\equiv_t$ are intimately related to distances between fuzzy sets.

The intuition behind this relation comes from the interpretation of $\equiv_t$ as a graded measure of equality of fuzzy sets and of their indistinguishability. The negation of such an indistinguishability relation $\equiv_t$ should hence be a kind of graded distinguishability and thus (perhaps) even a kind of "distance". And that really is the case.

In the following considerations we refer to the notion of a metric in $\mathbb{F}(\mathcal{X})$. As usual, a twoplace function ϱ from $\mathbb{F}(\mathcal{X})$ into the non-negative reals $\mathbb{R}^+$

is a *metric* iff for all $A, B, C \in I\!F(\mathcal{X})$ the following conditions hold true:

(M1)	$\varrho(A,B) = 0$ iff $A = B$,	(identity property)
(M2)	$\varrho(A,B) = \varrho(B,A)$,	(symmetry)
(M3)	$\varrho(A,C) + \varrho(C,B) \geq \varrho(A,B)$.	(triangle inequality)

Sometimes the identity condition (M1) is weakened to the condition

$$(\mathrm{M1}^p) \quad \varrho(A,A) = 0.$$

By a *pseudo-metric* ϱ then a function is meant that fulfills conditions (M1^p), (M2), (M3).

Definition 3.5 *For each left continuous t-norm* $\mathbf{t}$, *the binary* $\mathbf{t}$-distinguishability function $\varrho_{\mathbf{t}}$ *is that function on* $I\!F(\mathcal{X})$ *for which it is*

$$\varrho_{\mathbf{t}}(A,B) =_{def} 1 - [\![A \equiv_{\mathbf{t}} B]\!] \tag{3.16}$$

for all $A, B \in I\!F(\mathcal{X})$.

Obviously, for each $\mathbf{t}$-distinguishability function $\varrho_{\mathbf{t}}$ one has for all $A, B \in I\!F(\mathcal{X})$:

$$\varrho_{\mathbf{t}}(A,B) = [\![\neg(A \equiv_{\mathbf{t}} B)]\!].$$

For fuzzy sets $A, B \in \mathcal{X}$ in the case of the t-norm $\mathbf{t} = \mathbf{t}_G = \min$ one gets by simple calculations the corresponding distinguishability function $\varrho_G = \varrho_{\mathbf{t}_G}$

$$\begin{aligned} \varrho_G &= \sup_{\substack{x \in \mathcal{X} \\ [\![x\,\varepsilon\,A]\!] \neq [\![x\,\varepsilon\,B]\!]}} (1 - [\![x\,\varepsilon\,A \cap B]\!]) \\ &= \sup_{\substack{x \in \mathcal{X} \\ [\![x\,\varepsilon\,A]\!] \neq [\![x\,\varepsilon\,B]\!]}} ([\![x\,\varepsilon\,\complement(A \cap B)]\!]) \end{aligned}$$

and in the case of the t-norm $\mathbf{t} = \mathbf{t}_L$ one gets after some elementary transformations the distinguishability function $\varrho_L = \varrho_{\mathbf{t}_L}$ as

$$\begin{aligned} \varrho_L(A,B) = \min\Big\{1, &\max\{0, \sup_{x \in \mathcal{X}}([\![x\,\varepsilon\,A]\!] - [\![x\,\varepsilon\,B]\!])\} + \\ &+ \max\{0, \sup_{x \in \mathcal{X}}([\![x\,\varepsilon\,B]\!] - [\![x\,\varepsilon\,A]\!])\}\Big\}. \end{aligned}$$

This function $\varrho_L(A,B)$ is loosely related to the Čebyšev distance of the membership functions μ_A, μ_B defined as

$$d_C(\mu_A, \mu_B) = \sup_{x \in \mathcal{X}} |\mu_A(x) - \mu_B(x)|$$

in the sense that one always has

$$d_C(\mu_A, \mu_B) \le \varrho_L(A, B) \le 2 \cdot d_C(\mu_A, \mu_B)$$

and especially

$$\left(\models A \mathrel{\subseteqq_{t_L}} B\right) \quad \Rightarrow \quad \varrho_L(A, B) = d_C(\mu_A, \mu_B)$$

which means, using the original crisp implication relation $\subset$ for fuzzy sets as mentioned in (2.4),

$$A \subset B \quad \Rightarrow \quad \varrho_L(A, B) = d_C(\mu_A, \mu_B).$$

The problem now is to find a necessary and sufficient condition for $\mathbf{t}$ to yield via (3.16) a metric with properties (M1),..., (M3) as distinguishability function $\varrho_{\mathbf{t}}$.

As simple consequences of the definitions 2.4 of fuzzy singletons and 2.3 of fuzzified inclusion $\subseteqq_{\mathbf{t}}$ and identity $\equiv_{\mathbf{t}}$, first note that for all $a \in \mathcal{X}$ and $u, v \in [0,1]$ one has

$$[\![\langle\!\langle a \rangle\!\rangle_u \subseteqq_{\mathbf{t}} \langle\!\langle a \rangle\!\rangle_v]\!] = u \,\varphi_{\mathbf{t}}\, v = \sup\{w \mid u \,\mathbf{t}\, w \le v\} \tag{3.17}$$

and hence especially for $u = 1$:

$$[\![\langle\!\langle a \rangle\!\rangle_1 \subseteqq_{\mathbf{t}} \langle\!\langle a \rangle\!\rangle_v]\!] = v. \tag{3.18}$$

Now we can formulate and prove our characterization result.

Theorem 3.16 *Suppose* $\mathbf{t}$ *is a left continuous t-norm. Then the function* $\varrho_{\mathbf{t}}$ *of (3.16) is a metric in* $I\!F(\mathcal{X})$ *iff* $\mathbf{t} \geqq \mathbf{t}_L$*, i.e. iff for all* $u, v \in [0,1]$*:*

$$\max\{0, u + v - 1\} \le u \,\mathbf{t}\, v.$$

Proof. Obviously, (M2) holds true because $\equiv_{\mathbf{t}}$ is a symmetric relation. Condition (M1) by the corresponding definitions means

$$\varrho_{\mathbf{t}}(A, B) = 0 \;\Leftrightarrow\; [\![A \subseteqq_{\mathbf{t}} B]\!] = 1 \text{ and } [\![B \subseteqq_{\mathbf{t}} A]\!] = 1$$

for all $A, B \in I\!F(\mathcal{X})$. Additionally one has

$$[\![A \subseteqq_{\mathbf{t}} B]\!] = 1 \;\Leftrightarrow\; [\![x \,\varepsilon\, A \rightarrow_{\mathbf{t}} x \,\varepsilon\, B]\!] = 1 \text{ for all } x \in \mathcal{X}$$

and furthermore for all $u, v \in [0,1]$:

$$u \,\varphi_{\mathbf{t}}\, v = 1 \;\Leftrightarrow\; \sup\{w \mid u \,\mathbf{t}\, w \le v\} = 1.$$

Therefore proposition 1.6 (i) yields that

$$[\![A \subseteqq_t B]\!] \Leftrightarrow [\![x \,\varepsilon\, A]\!] \leq [\![x \,\varepsilon\, B]\!] = 1 \text{ for all } x \in \mathcal{X}$$

and hence also

$$\begin{aligned}[\![A \equiv_t B]\!] = 1 \quad &\Leftrightarrow \quad [\![x \,\varepsilon\, A]\!] = [\![x \,\varepsilon\, B]\!] \text{ for all } x \in \mathcal{X} \\ &\Leftrightarrow \quad A = B.\end{aligned}$$

To also get the triangle inequality (M3) one may start from the sup-t-transitivity of $\equiv_t$, cf. proposition 2.4 (iii), which can be written as

$$[\![A \equiv_t B \wedge_t B \equiv_t C]\!] \leq [\![A \equiv_t C]\!]$$

or equivalently as

$$\varrho_t(A, C) \leq 1 - [\![A \equiv_t B]\!] \,t\, [\![B \equiv_t C]\!]. \tag{3.19}$$

Then one has

$$\begin{aligned}\varrho_t(A, B) + \varrho_t(B, C) &= 1 - ([\![A \equiv_t B]\!] + [\![B \equiv_t C]\!] - 1) \\ &\geq 1 - t_L([\![A \equiv_t B]\!], [\![B \equiv_t C]\!]) \\ &\geq 1 - t([\![A \equiv_t B]\!], [\![B \equiv_t C]\!]) \\ &\geq \varrho_t(A, C)\end{aligned}$$

by (3.19), provided that $t \geqq t_L$.

Thus, if $t \geqq t_L$ holds true, ϱ_t is a metric.

Now assume that $t \geqq t_L$ is not the case, i.e. that there exist membership degrees $u_0, v_0 \in [0, 1]$ such that

$$u_0 \,t\, v_0 < u_0 \,t_L\, v_0 = \max\{0, u_0 = v_0 - 1\}.$$

Then consider points $a, b \in \mathcal{X}$ with $a \neq b$ and let be

$$\begin{aligned}A_0 &=_{def} \langle\!\langle a\rangle\!\rangle_1 \cup \langle\!\langle b\rangle\!\rangle_{v_0}, \\ B_0 &=_{def} \langle\!\langle a\rangle\!\rangle_{u_0} \cup \langle\!\langle b\rangle\!\rangle_{v_0}, \\ C_0 &=_{def} \langle\!\langle a\rangle\!\rangle_{u_0} \cup \langle\!\langle b\rangle\!\rangle_1\end{aligned}$$

unions of suitable fuzzy singletons of a and b. By straightforward calculations one gets via (3.17), (3.18):

$$\begin{aligned}[\![A_0 \equiv_t B_0]\!] &= [\![\langle\!\langle a\rangle\!\rangle_1 \subseteqq_t \langle\!\langle a\rangle\!\rangle_{u_0}]\!] = u_0, \\ [\![B_0 \equiv_t C_0]\!] &= [\![\langle\!\langle b\rangle\!\rangle_1 \subseteqq_t \langle\!\langle b\rangle\!\rangle_{v_0}]\!] = v_0, \\ [\![A_0 \equiv_t C_0]\!] &= [\![\langle\!\langle a\rangle\!\rangle_1 \subseteqq_t \langle\!\langle a\rangle\!\rangle_{u_0}]\!] \,t\, [\![\langle\!\langle b\rangle\!\rangle_1 \subseteqq_t \langle\!\langle b\rangle\!\rangle_{v_0}]\!] = u_0 \,t\, v_0.\end{aligned}$$

Hence one has that

$$\begin{aligned}&[\![A_0 \equiv_{\boldsymbol{t}} B_0 \wedge_{\boldsymbol{t}} B_0 \equiv_{\boldsymbol{t}} C_0]\!] = [\![A_0 \equiv_{\boldsymbol{t}} C_0]\!] = u_0 \, \boldsymbol{t} \, v_0 \\ &\quad < u_0 \, \boldsymbol{t}_L \, v_0 = [\![A_0 \equiv_{\boldsymbol{t}} B_0 \,\&\, b_0 \equiv_{\boldsymbol{t}} C_0]\!]\end{aligned}$$

and therefore

$$\begin{aligned}\varrho_{\boldsymbol{t}}(A_0, B_0) + \varrho_{\boldsymbol{t}}(B_0, C_0) &= 1 - ([\![A_0 \equiv_{\boldsymbol{t}} B_0]\!] + [\![B_0 \equiv_{\boldsymbol{t}} C_0]\!] - 1) \\ &= 1 - [\![A_0 \equiv_{\boldsymbol{t}} B_0 \,\&\, B_0 \equiv_{\boldsymbol{t}} C_0]\!] \\ &< 1 - [\![A_0 \equiv_{\boldsymbol{t}} C_0]\!] \;=\; \varrho_{\boldsymbol{t}}(A_0, C_0),\end{aligned}$$

i.e. the triangle inequality does not hold true in this case.

Thus, if $\boldsymbol{t} \geqq \boldsymbol{t}_L$ does not hold true for $\varrho_{\boldsymbol{t}}$, the triangle inequality fails and hence $\varrho_{\boldsymbol{t}}$ is not a metric. QED

A close look at the way our last theorem 3.16 was stated and proven shows that it is enough to have a function φ introduced by definition (1.19) for a given t-norm $\boldsymbol{t}$. That means to some extent we can avoid supposing that this function φ is a Φ-operator.

Formally, as we already mentioned in chapter 1, this definition does not need the assumption of the left continuity of $\boldsymbol{t}$. Thus, perhaps for the present purposes of getting a metric for fuzzy sets, this assumption is not necessary or can be weakened.

Indeed, our arguments to establish that $\varrho_{\boldsymbol{t}}$ has properties (M2) and (M3) do not use the left continuity of $\boldsymbol{t}$. Thus, the crucial point is the proof of (M1). And for that case the following proposition holds true.

Proposition 3.17 *For $\varrho_{\boldsymbol{t}}(A,B) =_{def} 1 - [\![A \equiv_{\boldsymbol{t}} B]\!]$ there holds true*

$$\varrho_{\boldsymbol{t}}(A,B) = 0 \;\Leftrightarrow\; A = B \;\; \textit{for all } A, B \in I\!F(\mathcal{X})$$

iff each function $\boldsymbol{t}_u =_{def} \lambda v(u \, \boldsymbol{t} \, v)$, $u \in [0,1]$ is (left) continuous at $v = 1$.

Proof. By the corresponding definitions one has

$$\begin{aligned}\varrho_{\boldsymbol{t}}(A,B) = 0 \;&\Leftrightarrow\; [\![A \equiv_{\boldsymbol{t}} B]\!] = 1 \\ &\Leftrightarrow\; [\![A \subseteqq_{\boldsymbol{t}} B \wedge_{\boldsymbol{t}} B \subseteqq_{\boldsymbol{t}} A]\!] = 1 \\ &\Leftrightarrow\; [\![A \subseteqq_{\boldsymbol{t}} B]\!] = 1 \;\text{ and }\; [\![B \subseteqq_{\boldsymbol{t}} A]\!] = 1\end{aligned}$$

and furthermore

$$\begin{aligned}[\![A \subseteqq_{\boldsymbol{t}} B]\!] = 1 \;&\Leftrightarrow\; \inf_{x \in \mathcal{X}} [\![x \,\varepsilon\, A \rightarrow_{\boldsymbol{t}} x \,\varepsilon\, B]\!] = 1 \\ &\Leftrightarrow\; [\![x \,\varepsilon\, A \rightarrow_{\boldsymbol{t}} x \,\varepsilon\, B]\!] = 1 \;\text{ for all } x \in \mathcal{X}.\end{aligned}$$

No left continuity of $\boldsymbol{t}$ is used here. Using only (1.19) now one gets

$$[\![u \rightarrow_{\boldsymbol{t}} v]\!] = 1 \Leftrightarrow \sup\{w \mid u\,\boldsymbol{t}\,w \leq v\} = 1$$

for all $u, v \in [0,1]$. All together that means

$$\begin{aligned} &\varrho_{\boldsymbol{t}}(A,B) = 0 \Leftrightarrow \\ &\qquad [\![x \,\varepsilon\, A \rightarrow_{\boldsymbol{t}} x \,\varepsilon\, B]\!] \text{ and } [\![x \,\varepsilon\, B \rightarrow_{\boldsymbol{t}} x \,\varepsilon\, A]\!] = 1 \text{ for all } x \in \mathcal{X} \end{aligned} \tag{3.20}$$

and the equivalence of the following two conditions:

$$\varrho_{\boldsymbol{t}}(A,B) = 0 \Leftrightarrow A = B$$

and

$$\begin{aligned} &A = B \Leftrightarrow \\ &\qquad [\![x \,\varepsilon\, A \rightarrow_{\boldsymbol{t}} x \,\varepsilon\, B]\!] = 1 \text{ and } [\![x \,\varepsilon\, B \rightarrow_{\boldsymbol{t}} x \,\varepsilon\, A]\!] = 1 \text{ for all } x \in \mathcal{X}. \end{aligned} \tag{3.21}$$

It is obvious that

$$u \leq v \Rightarrow [\![u\,\varphi_{\boldsymbol{t}}\,v]\!] = 1$$

holds true also without reference to the left continuity of $\boldsymbol{t}$. Hence part ($\Rightarrow$) of (3.21) is obvious too. But what about the reverse implication

$$[\![u\,\varphi_{\boldsymbol{t}}\,v]\!] = 1 \Rightarrow u \leq v,$$

i.e. what about

$$u > v \Rightarrow [\![u\,\varphi_{\boldsymbol{t}}\,v]\!] < 1?$$

Assume that $u_0 > v_0$ and $[\![u_0\,\varphi_{\boldsymbol{t}}\,v_0]\!] = \sup\{w \mid u_o\,\boldsymbol{t}\,w \leq v_0\} < 1$ holds true for some $u_0, v_0 \in [0,1]$. Then there is a $w_0 < 1$ such that $u_0\,\boldsymbol{t}\,w_0 > v_0$ and, of course, $v_0 < 1$ hold true. Hence, if $u_0 > v_0$ and $[\![u_0\,\varphi_{\boldsymbol{t}}\,v_0]\!] = 1$ hold true, then the function $\boldsymbol{t}_{u_0} = \boldsymbol{\lambda} v(u_0\,\boldsymbol{t}\,v)$ is not left continuous at point $v = 1$. Indeed otherwise one would have $\sup\{w \mid u_0\,\boldsymbol{t}\,w \leq v_0\} = 1$ and therefore

$$\begin{aligned} v_0 &\geq \sup\{u_0\,\boldsymbol{t}\,w \mid u_0\,\boldsymbol{t}\,w \leq v_0\} \\ &= u_0\,\boldsymbol{t}\,\sup\{w \mid u_0\,\boldsymbol{t}\,w \leq v_0\} = u_0\,\boldsymbol{t}\,1 = u_0. \end{aligned}$$

On the other hand, assuming that for each $u \in [0,1]$ the function $\boldsymbol{t}_u = \boldsymbol{\lambda} v(u\,\boldsymbol{t}\,v)$ is left continuous at $v = 1$ gives for all $u_0, v_0 \in [0,1]$:

$$\begin{aligned} u_0\,\varphi_{\boldsymbol{t}}\,v_0 &= \sup\{w \mid u_0\,\boldsymbol{t}\,w \leq v_0\} = 1 \\ &\Rightarrow u_0 = u_0\,\boldsymbol{t}\,1 = u + 0\,\boldsymbol{t}\,\sup\{w \mid u_0\,\boldsymbol{t}\,w \leq v_0\} \\ &\qquad = \sup\{u_0\,\boldsymbol{t}\,w \mid u_0\,\boldsymbol{t}\,w \leq v_0\} \leq v_0, \end{aligned}$$

i.e.

$$[\![u_0 \varphi_t v_0]\!] = 1 \Rightarrow u_0 \leq v_0.$$

Hence assuming that all t_u are left continuous at $v = 1$ gives

$$[\![x \,\varepsilon\, A \rightarrow_t x \,\varepsilon\, B]\!] = 1 \text{ and } [\![x \,\varepsilon\, B \rightarrow_t x \,\varepsilon\, A]\!] = 1$$
$$\Rightarrow [\![x \,\varepsilon\, A]\!] \leq [\![x \,\varepsilon\, B]\!] \text{ and } [\![x \,\varepsilon\, B]\!] \leq [\![x \,\varepsilon\, A]\!]$$

for all $x \in \mathcal{X}$ and thus also the ($\Leftarrow$)-part of (3.21). QED

This proposition therefore shows that the continuity assumption in the theorem 3.16 can be weakened.

Furthermore, by the reflexivity of $\equiv_t$ it is obvious that condition (M1^p), which instead of (M1) is used to define pseudo-metrics, always holds true for ϱ_t. Thus the following results hold true too.

Corollary 3.18 *ϱ_t is a metric iff $t \geqq t_L$ and each one of the functions $t_u = \lambda v(u\,t\,v)$ is left continuous at the point $v = 1$.*

Corollary 3.19 *ϱ_t is a pseudo-metric iff $t \geqq t_L$.*

Unfortunately, however, it is not clear whether these corollaries are as interesting as the theorem. The crucial point here is that in the case of a t-norm t which is not left continuous, for φ_t the property

$$u\,t\,(u \varphi_t v) \leq v \quad \text{for all } u, v \in [0,1] \tag{3.22}$$

and thus (Φ2) fails, as was shown in the proof of the Existence Lemma. But (Φ2), one of the properties which in chapter 1 were used to define Φ-operators, means that the generalized conjunction $\wedge_t$ and the generalized implication $\rightarrow_t$ together fulfill a kind of generalized modus ponens: the truth degree of a well-formed formula H_2 is never smaller than the truth degree of the $\wedge_t$-conjunction of formulas H_1 and $H_1 \rightarrow_t H_2$.

So we have the situation that the greater generality of those corollaries with respect to theorem 3.16 rests on the assumption that the modus ponens condition (3.22) is not essential for the context the corollaries are used in.

Of course, the distances ϱ_t are *bounded*: $\varrho_t(A, B) \leq 1$ always applies.

For each fuzzy equation (3.9): $\Theta(X, A) = B$ and each fuzzy relation R_0 thus the truth degree

$$\xi_0(R_0) =_{def} [\![\Theta(R_0, A) \equiv_t B]\!]$$

not only evaluates the truth of the sentence "R_0 *is a solution of equation* $\Theta(X, A) = B$", i.e. measures the degree to which R_0 is a solution of the equation, but at the same time the degree

$$\eta_0(R_0) =_{def} 1 - \xi_0(R_0),$$

i.e. the degree

$$\eta_0(R_0) = \varrho_t(\Theta(R_0, A), B) = [\![\neg(\Theta(X, A) \equiv_t B)]\!]$$

evaluates the distance in between, i.e. the difference of the fuzzy sets $\Theta(R_0, A)$ and B, and thus characterizes how "far" R_0 is from a true solution.

For systems (3.10) of fuzzy equations the situation is the same. For any fuzzy relation R_0 the degree

$$\xi(R_0) =_{def} [\![\prod_{i=1}^{n}(\Theta(R_0, A_i) \equiv_t B_i)]\!]$$

evaluates to what extent R_0 is a solution of the system (3.10). Hence also in the present case the degree

$$\eta(R_0) =_{def} 1 - \xi(R_0) = [\![\neg(\prod_{i=1}^{n}(\Theta(R_0, A_i) \equiv_t B_i))]\!]$$

characterizes how "far" R_0 is from a true solution to the system of fuzzy equations.

The reference to a true solution is not essential for these considerations. Thus the degrees $\eta_0(R_0)$ and $\eta(R_0)$ can be considered as measuring the *distance* between the left-hand side(s) of (3.9) and (3.10) "produced" by taking $X = R_0$ and their "ideal" values, i.e. the right-hand sides of (3.9) and (3.10) respectively. Vanishing distances give $\eta_0(R_0) = 0$ or $\eta(R_0) = 0$ and thus $\xi_0(R_0) = 1$ or $\xi(R_0) = 1$. On the other hand, the worst cases are $\eta_0(R_0) = 1$ and $\eta(R_0) = 1$ and thus $\xi_0(R_0) = 0$ and $\xi(R_0) = 0$.

That means that the degrees $\xi_0(R_0)$ in the case of a single equation (3.9) and $\xi(R_0)$ in the case of a system (3.10) of fuzzy equations measure a kind of *approximation quality* of R_0 with respect to some (possibly nonexistent) ideal solution. Therefore the solvability degrees

$$\xi_0 = \sup_R \xi_0(R) \qquad \text{and} \qquad \xi = \sup_R \xi(R),$$

with the supremum in both cases having taken over all possible fuzzy relations R, are upper bounds for the quality of approximate solvability of

(3.9) and (3.10) and hence the ***solvability degrees measure the best possible approximate solvability*** of (systems of) fuzzy equations in sense of the distances ϱ_t.

The situation is especially simple in cases of single fuzzy equations. Reconsidering from the present point of view our earlier results in theorems 3.4 and 3.7 we already determined there the best possible approximate solutions. Those results, under this new perspective, are once again presented in Table 3.1.

fuzzy equation	best possible approximate solution
$R''X = B$	$R \downarrow B$
$X''A = B$	$A \rhd_t B$
$R \circ_t X = S$	$R^{-1} \diamond_t S$
$X \circ_t R = S$	$(R \diamond_t S^{-1})^{-1}$
$A *_t X = B$	$B \tilde{*}_t A$

Table 3.1: Approximate solutions of some fuzzy equations

For systems of fuzzy equations the situation is more difficult. Without restrictions concerning the t-norm which always is considered to be given, theorem 3.9 only gives inequalities. But if one restricts the considerations to the t-norm $t = t_G = \min$, then one can extend the results of corollary 3.10 to all types of equations in Table 3.1 in an (almost) uniform way and has the following

Proposition 3.20 *Suppose* $t = t_G = \min$. *For each type of fuzzy equations in Table 3.1 then there exist best possible approximate solutions to systems of such fuzzy equations; and in each case the intersection of all best possible approximate solutions, as given in Table 3.1, of the single equations of the system is such a best possible approximate solution to the whole system.*

Surely, in the definitions of solvability degrees ξ, ξ_0 also other kinds of "distances" of fuzzy sets apart from our $\equiv_t$-based (pseudo-) metrics ϱ_t could have been used. We have not discussed such possibilities so far in detail. The reference to $\equiv_t$ has the advantage that in the proofs of characterizations of the solvability degrees (3.11), (3.12), our above methods are available which are quite straightforward generalizations (to many-valued logic) of usual methods from classical set theory and logic.

As proved in theorem 3.4 for lower semicontinuous t-norms $\boldsymbol{t}$, the solvability degree (3.11) can be characterized as the value

$$\xi_0 = [\![(A \rhd_{\boldsymbol{t}} B)'' A \equiv_{\boldsymbol{t}} B]\!]$$

where the fuzzy relation $A \rhd_{\boldsymbol{t}} B$ is defined according to definition 3.2.

From proposition 3.1 (ii) we have for the solvability degree ξ_0 of the equation (3.2)

$$\xi_0 = \bigwedge_{y \in \mathcal{Y}} ([\![y \varepsilon B]\!] \varphi_{\boldsymbol{t}} \bigvee_{x \in \mathcal{X}} ([\![x \varepsilon A]\!] \, \boldsymbol{t} \, ([\![x \varepsilon A]\!] \varphi_{\boldsymbol{t}} [\![y \varepsilon B]\!]))) \tag{3.23}$$

by definition 2.3 and well known properties of Φ-operators and t-norms.

The situation is a bit more complicated for systems (3.4) of fuzzy relation equations: in theorem 3.9, only lower and upper bounds for the solvability degree (3.12) could be given in general, and a full characterization only in the case that $\boldsymbol{t}$ = min. For systems (3.4) of fuzzy relation equations the result was

$$\xi \geq \mathop{\mathbf{T}}_{i=1}^{n} [\![((\bigcap_{j=1}^{n}(A_j \rhd_{\boldsymbol{t}} B_j))'' A_i \equiv_{\boldsymbol{t}} B_i)]\!] \geq \mathop{\mathbf{T}}_{i=1}^{n} \xi$$

for lower semicontinuous t-norms $\boldsymbol{t}$. Hence, by definition 3.2 and the monotonicity of $\circ_{\boldsymbol{t}}$ with respect to the inclusion of fuzzy sets we obtain

$$\begin{aligned} \xi \;\geq\; & \mathop{\mathbf{T}}_{i=1}^{n} \bigwedge_{y \in \mathcal{Y}} ([\![y \varepsilon B_i]\!] \varphi_{\boldsymbol{t}} \bigvee_{x \in \mathcal{X}} ([\![x \varepsilon A_i]\!] \, \boldsymbol{t} \bigwedge_{1 \leq j \leq n} ([\![x \varepsilon A_j]\!] \varphi_{\boldsymbol{t}} [\![y \varepsilon B_j]\!]))) \\ & \geq \mathop{\mathbf{T}}_{i=1}^{n} \xi. \end{aligned} \tag{3.24}$$

To get simpler formulas we now suppose that our t-norms $\boldsymbol{t}$ are continuous. Then using proposition 1.7 we get a much simpler formulation of the solvability degree (3.11) than earlier in (3.23). Remember for this that it is

$$\operatorname{hgt}(A) = \bigvee_{x \in \mathcal{X}} [\![x \varepsilon A]\!] = [\![\exists x (x \varepsilon A)]\!]. \tag{3.25}$$

Proposition 3.21 *For continuous t-norms* $\boldsymbol{t}$ *one has for the solvability degree* ξ_0 *of fuzzy relation equation (3.2)*

$$\xi_0 = \operatorname{hgt}(B) \, \varphi_{\boldsymbol{t}} \operatorname{hgt}(A).$$

Proof. From (3.23) and the monotonicity of φ_t in the second argument we get

$$\xi_0 = \bigwedge_{y\in\mathcal{Y}} \bigvee_{x\in\mathcal{X}} ([\![y \,\varepsilon\, B]\!] \,\varphi_t\, ([\![x \,\varepsilon\, A]\!] \,t\, ([\![x \,\varepsilon\, A]\!] \,\varphi_t\, [\![y \,\varepsilon\, B]\!])))$$

and hence by proposition 1.7 (ii) and monotonicity

$$\xi_0 = \bigwedge_{y\in\mathcal{Y}} ([\![y \,\varepsilon\, B]\!] \,\varphi_t \bigvee_{x\in\mathcal{X}} [\![x \,\varepsilon\, A]\!]).$$

Now, using results of chapter 1 we get by continuity of t

$$\xi_0 = (\bigvee_{y\in\mathcal{Y}} [\![y \,\varepsilon\, B]\!]) \,\varphi_t\, (\bigvee_{x\in\mathcal{X}} [\![x \,\varepsilon\, A]\!])$$

which is what is to be proved according to (3.25). QED

Corollary 3.22 *For each continuous t-norm* t *for the solvability degree* ξ_0 *of equation (3.2) there hold*

(*i*) $\quad \mathrm{hgt}\,(A) \le \xi_0 \le 1,$

(*ii*) $\quad \mathrm{hgt}\,(A) = 1 \Rightarrow \xi_0 = 1,$

(*iii*) $\quad \mathrm{hgt}\,(B) = 1 \Rightarrow \xi_0 = \mathrm{hgt}\,(A).$

Proof. Obvious.

Remember inequalities (3.24). Put simply $j = i$ in the minimum on all $1 \le j \le n$. From monotonicity of t and of φ_t in the second argument then we get for the solvability degree ξ of (3.12)

$$\mathop{\mathbf{T}}_{i=1}^{n} \xi \le \mathop{\mathbf{T}}_{i=1}^{n} \bigwedge_{y\in\mathcal{Y}} ([\![y \,\varepsilon\, B_i]\!] \,\varphi_t \bigvee_{x\in\mathcal{X}} ([\![x \,\varepsilon\, A_i]\!] \,t\, ([\![x \,\varepsilon\, A_i]\!] \,\varphi_t\, [\![y \,\varepsilon\, B_i]\!])))$$

which by the proof of proposition 3.21 becomes

$$\mathop{\mathbf{T}}_{i=1}^{n} \xi \le \mathop{\mathbf{T}}_{i=1}^{n} (\mathrm{hgt}\,(B_i) \,\varphi_t\, \mathrm{hgt}\,(A_i)).$$

Hence we have proved the following proposition.

Proposition 3.23 *Let* $\mathbf{t}$ *be a continuous t-norm. Denote by* ξ_i *the solvability degree of the i-th equation of system (3.4). Then for the solvability degree* ξ *of the system (3.4) one has*

$$\mathop{\mathbf{T}}_{i=1}^{n} \xi \leq \mathop{\mathbf{T}}_{i=1}^{n} \xi_i.$$

Furthermore, from (3.24) and corresponding results of chapter 1 we get immediately for this solvability degree

$$\xi \geq \mathop{\mathbf{T}}_{i=1}^{n} \bigwedge_{y \in \mathcal{Y}} \bigvee_{x \in \mathcal{X}} \bigwedge_{1 \leq j \leq n} ([\![y \varepsilon B_i]\!] \varphi_{\mathbf{t}} ([\![x \varepsilon A_i]\!] \, \mathbf{t} ([\![x \varepsilon A_j]\!] \varphi_{\mathbf{t}} [\![y \varepsilon B_j]\!]))).$$

Now we can use $u \varphi_{\mathbf{t}} v \geq v$ to reduce the term $[\![x \varepsilon A_j]\!] \varphi_{\mathbf{t}} [\![y \varepsilon B_j(y)]\!]$ to $[\![y \varepsilon B_j]\!]$ and "move back" the supremum on $x \in \mathcal{X}$ to the second argument of the Φ-operator and finally to the first argument of the following t-norm term. This gives

Proposition 3.24 *Let* $\mathbf{t}$ *be a continuous t-norm. Then,*

$$\xi \geq \mathop{\mathbf{T}}_{i=1}^{n} \bigwedge_{y \in \mathcal{Y}} ([\![y \varepsilon B_i]\!] \varphi_{\mathbf{t}} (\operatorname{hgt}(A_i) \, \mathbf{t} ([\![y \varepsilon \bigcap_{j=1}^{n} B_j]\!])))$$

holds for the solvability degree ξ *of system (3.4) of fuzzy equations.*

Unfortunately, hence, for systems of fuzzy relation equations we get only bounds for the solvability degree, while for single fuzzy relation equations we obtained a full characterization.

Another difference seems worth mentioning for applications. If we assume that the input data are described by normal fuzzy sets, i.e. by such ones whose height is $=1$, then proposition 3.21 always gives solvability degree $=1$ for the single fuzzy relation equation. Therefore proposition 3.23 also gives only $\xi \leq 1$ for the solvability degree of system (3.4). But from proposition 3.24 we get a simpler formula for a lower bound of ξ in this case. To write down this formula, we have to use the fuzzified inclusion $\subseteqq_{\mathbf{t}}$.

Corollary 3.25 *Let* $\mathbf{t}$ *be a continuous t-norm. Suppose that the fuzzy sets* A_i *in the system (3.4) are normal fuzzy sets for all indices i. Then*

$$\xi \geq \mathop{\mathbf{T}}_{i=1}^{n} [\![B_i \subseteqq_{\mathbf{t}} \bigcap_{j=1}^{n} B_j]\!].$$

Proof. Straightforward.

One has even better results for the t-norm min. In this special case obviously $\mathop{\mathbf{T}}_{i=1}^{n} \ldots = \min_{1\le i\le n} \ldots$ and hence (3.24) is an equality, i.e. it gives a characterization of the solvability degree of system (3.4) in terms of the input and output fuzzy sets. Nevertheless, even the bounds which result from propositions 3.23 and 3.24 are interesting in this case.

Corollary 3.26 *Consider the t-norm min. Then*

$$\min\{\bigwedge_{1\le i\le n} \xi_i, [\![\bigcup_{i=1}^{n} B_i \subseteq_t \bigcap_{j=1}^{n} B_j]\!]\} \le \xi \le \bigwedge_{1\le i\le n} \xi_i.$$

Proof. For $t = t_G = \min$ the corresponding Φ-operator φ_G (cf. table 2.3) distributes over this t-norm. Hence we get from proposition 3.24 now

$$\xi \ge \bigwedge_{1\le i\le n} \bigwedge_{y\in\mathcal{Y}} \min\{[\![y \,\varepsilon\, B_i]\!] \,\varphi_G\, \mathrm{hgt}\,(A_i), [\![y \,\varepsilon\, B_i]\!] \varphi_G [\![y \,\varepsilon \bigcap_{j=1}^{n} B_j]\!]\}.$$

From this, elementary calculations give

$$\xi \ge \min\{\bigwedge_{1\le i\le n} (\mathrm{hgt}\,(B_i)\,\varphi_G\,\mathrm{hgt}\,(A_i)), \bigwedge_{y\in\mathcal{Y}} ([\![y \,\varepsilon \bigcup_{i=1}^{n} B_i]\!] \,\varphi_G\, [\![y \,\varepsilon \bigcap_{j=1}^{n} B_j]\!])\}$$

which proves the lower bound of ξ by proposition 3.21 and definition 2.3. The upper bound immediately results from proposition 3.23. QED

3.6 Towards more difficult equations

For this section we adopt the notation $A \circ_t R = B$ instead of our earlier $R''A = B$, and additionally we use ZADEH's original membership function notation $\mu_A(x), \ldots$ instead of our $[\![x \,\varepsilon\, A]\!]$. The reason is that in this way a simpler comparison can be made with parts of the research literature in the engineering field.

Types of fuzzy relational equations studied so far and which are of special significance for applications in the field of fuzzy modelling include the following:

- fuzzy relational equations with sup-t-composition

$$B = A \circ_t R, \tag{3.26}$$
$$\mu_B(y) = \sup_{x \in \mathcal{X}}(\mu_A(x)\, t\, \mu_R(x,y)), \tag{3.27}$$

  which is our type (3.2) of equations and which is strongly connected with the "compositional rule of inference" of ZADEH (1973), cf. also chapter 4;

- fuzzy relational equations with inf-s-composition

$$B = A \diamond_t R, \tag{3.28}$$
$$\mu_B(y) = \inf_{x \in \mathcal{X}}(\mu_A(x)\, s_t\, \mu_R(x,y)), \tag{3.29}$$

  which in a suitable sense is dual to the foregoing case;

- the adjoint fuzzy relational equations of PEDRYCZ (1985) with inf-φ-composition

$$B = A \rhd_t R, \tag{3.30}$$
$$\mu_B(y) = \inf_{x \in \mathcal{X}}(\mu_A(x)\, \varphi_t\, \mu_R(x,y)). \tag{3.31}$$

Besides these types, which might be viewed as beingof a basic nature, one can consider some types of a more complex form which sometimes are formed on the basis of the types given above. Of such possibilities we recall two:

- a convex combination of equations of types (3.26) and (3.28) as proposed in OHSATO/SEKIGUCHI (1983)

$$B = \lambda \cdot (A \circ_t R_1) + (1-\lambda)\cdot(A \diamond_{s_{t_1}} R_2), \tag{3.32}$$
$$\mu_B(y) = \lambda(y)\cdot(\sup_{x \in \mathcal{X}}(\mu_A(x)\, s_t\, \mu_{R_1}(x,y))) + $$
$$+(1-\lambda(y))\cdot \inf_{x \in \mathcal{X}}(\mu_A(x)\, s_{t_1}\, \mu_{R_2}(x,y)) \tag{3.33}$$

  with $\lambda : \mathcal{Y} \to [0,1]$ and the t-norm t, as well as the t-conorm s_{t_1}, in OHSATO/SEKIGUCHI (1983) taken as max, min only;

- fuzzy relational equations with an equality operator as discussed in DINOLA/PEDRYCZ/SESSA (1988)

$$B = A \bowtie_t R, \tag{3.34}$$
$$\mu_B(y) = \sup_{x \in \mathcal{X}}(\mu_A(x)\, \varphi_t\, \mu_R(x,y) \wedge (\mu_R(x,y)\, \varphi_t\, \mu_A(x))), \tag{3.35}$$

For all the basic forms of equations (3.28) – (3.31) the family of solutions, if solutions do exist at all, has been characterized and their extremal (maximal or minimal) elements have been obtained. Moreover, for systems

$$B_i = A_i \circ_{\mathbf{t}} R, \quad i = 1, \ldots, n \tag{3.36}$$

$$B_i = A_i \diamond_{\mathbf{t}} R, \quad i = 1, \ldots, n \tag{3.37}$$

$$B_i = A_i \rhd_{\mathbf{t}} R, \quad i = 1, \ldots, n \tag{3.38}$$

of such equations the relevant results are also available – mainly under the condition that they do have an exact solution.

To have an overall picture of the results they are collected in Table 3.2.

Concerning the notation used in this table we have to add two explanations. First, by $\boldsymbol{R}', \boldsymbol{R}'', \boldsymbol{R}'''$ we denote the sets of solutions of the systems (3.36), (3.37), (3.38) respectively of fuzzy relational equations. Secondly, as a dual to the Φ-operator (1.19) we use the sometimes so-called β-operator defined for all $u, v \in [0,1]$ as:

$$u \, \beta_{\mathbf{t}} \, v =_{def} \inf\{w \in [0,1] \mid u \, \mathbf{s}_{\mathbf{t}} \, w \geq v\}; \tag{3.39}$$

and in an analogous manner as the operator $\rhd_{\mathbf{t}}$ was introduced in definition 3.2 with reference to the implication connective $\rightarrow_{\mathbf{t}}$, i.e. to the Φ-operator $\varphi_{\mathbf{t}}$, now a "dual" operator $\lhd_{\mathbf{t}}$ for fuzzy sets is introduced such that $C := A \lhd_{\mathbf{t}} B$ is characterized by the membership function

$$C := A \lhd_{\mathbf{t}} B : \qquad \mu_C(x,y) =_{def} \mu_A(x) \, \beta_{\mathbf{t}} \, \mu_B(y).$$

At present however, for the general case it seems quite difficult to give simple and easy-to-check conditions for the solvability of a system of equations. The discussion in chapter 4 concerning sufficiency conditions for the non-interactivity of systems of control rules gives an indication of the problems one is confronted with in looking for *simple* conditions here. This was one of the reasons for the discussion of degrees of solvability for example in GOTTWALD (1986) and in this chapter. Actually we will present only some conditions for the solvability of single equations; cf. Table 3.3. The case of systems of such fuzzy relational equations has not yet been considered in detail.

As it becomes quite clear from later discussions concerning applications to fuzzy control and fuzzy modelling in general, the results for example of Table 3.2 have a significant value only in the case that solutions really exist, i.e. that not only "approximate" solutions (in some suitable sense of that word) exist. If this true solvability is not the case – and it is this more

Type of equation	Interpretation	Solution[3] to a single equation	Solution[3] to a system of equations
$B = A \circ_t R$, sup-t-composition	the (fuzzified) full image of a fuzzy set under a fuzzy relation *or* for each point b of B there is a point a of A such that b can be reached from point a	$\hat{R} = A \rhd_t B$ $\hat{R} = \sup \boldsymbol{R}'$	$\hat{R} = \bigcap_{i=1}^{n}(A_i \rhd_t B_i)$ $\hat{R} = \bigcap_{i=1}^{n} \sup \boldsymbol{R}'_i$
$B = A \diamond_t R$, inf-s-composition	dual to the first case in sense that $B = A \diamond_t R$ iff $\bar{B} = \bar{A} \circ_t \bar{R}$	$\check{R} = A \lhd_t B$ $\check{R} = \inf \boldsymbol{R}''$	$\check{R} = \bigcup_{i=1}^{n}(A_i \lhd_t B_i)$ $\check{R} = \bigcup_{i=1}^{n} \inf \boldsymbol{R}''$
$B = A \rhd_t R$, inf-φ-composition	each point b of B can be reached from every point a of A	$\tilde{R} = A \times_t B$ $\tilde{R} = \inf \boldsymbol{R}'''$	$\tilde{R} = \bigcup_{i=1}^{n}(A_i \times_t B_i)$ $\tilde{R} = \bigcup_{i=1}^{n} \inf \boldsymbol{R}'''$

Table 3.2: Basic types of fuzzy relation equations for control applications

uncomfortable situation one usually meets in practice – then because of the (present) lack of an extended mathematical theory of such (systems of) equations, the user has to think about other ways of overcoming the problem of the nonexistence of (true) solutions.

A simple, and perhaps for the practitioner the most obvious, way out is to use the formulas which describe solutions – in the case of solvability – even if a solution to the system of fuzzy relation equations to be considered does not exist – and then to check the quality of the "approximate solution" derived in this way.

But having taken this point of view one can move one step further: instead of having proven a formula to give a solution in the case of solvability, one can start from a formula which one guesses to describe a solution – of course, if there are some acceptable reasons for such a guess. And, indeed, for some classes of fuzzy relational equations such acceptable guesses are available. To present some basic ones let us distinguish for relational equations $\Theta(R, A) = B$ two different types.

[3] in case of solvability

Type of equation	Condition
$B = A \circ_t R$	$\bigwedge_{y\in\mathcal{Y}} \bigvee_{x\in\mathcal{X}} (\mu_A(x) \geq \mu_B(y))$
$B = A \diamond_t R$	$\bigwedge_{y\in\mathcal{Y}} \bigvee_{x\in\mathcal{X}} (\mu_A(x) \leq \mu_B(y))$
$B = A \rhd_t R$	no restrictive condition in the case that t is strictly monotonic in both arguments

Table 3.3: Necessary and sufficient conditions that a single equation has a solution.

Definition 3.6 *A fuzzy equation $\Theta(R, A) = B$ will be said to be of* sup-type *in the case that one has*

$$\mu_B(y) = \mu_\Theta(A, R)(y) = \sup_{x\in\mathcal{X}} \Gamma(\mu_A(x), \mu_R(x, y)) \tag{3.40}$$

where the term Γ is built up using the membership degrees $\mu_A(a)$, $\mu_R(a, b)$ and combining them for example by a t-norm, a Φ-operator or some suitable other kinds of "simple" operators; and such an equation will be said to be of inf-type *in the case that one has*

$$\mu_B(y) = \mu_\Theta(A, R)(y) = \inf_{x\in\mathcal{X}} \Gamma(\mu_A(x), \mu_R(x, y)) \tag{3.41}$$

with the term Γ chosen accordingly.

To show the influence of this distinction on the structure of the set of solutions of these equations we will consider the following facts which should be taken into account for the discussions of true and of approximate solutions.

Fact 1: The union of any two solutions of a fuzzy relational equation of sup-type is again a solution to this equation, and hence this equation has a greatest solution.

Fact 2: The intersection of any two solutions of a fuzzy relational equation of inf-type is again a solution to this equation, and hence this equation has a smallest solution.

Fact 3: Any system of fuzzy relational equations of sup-type has in the case of solvability as the greatest solution the intersection of all the greatest solutions of its single equations.

Fact 4: Any system of fuzzy relational equations of inf-type has in the case of solvability as the smallest solution the union of all the smallest solutions of its single equations.

It is interesting to mention that these types of behaviour can also be found with some mixed types of fuzzy relational equations. To look at an example we refer to the convex combination form (3.32), (3.33) of fuzzy relational equations discussed by OHSATO/SEKIGUCHI (1983,1985) for $A \circ_t R_1$ as sup-min-composition and $A \diamond_t R_2$ as inf-max-composition. Indeed, in OHSATO/SEKIGUCHI (1985) a solvability behaviour of such equations is proven which combines Facts 1 and 2: the type (3.32) of equations has an extremal solution $(\check{R}_1, \hat{R}_2)$ in the sense that $R_1' \subseteq \check{R}_1$ and $R_2' \subseteq \hat{R}_2$ for all solutions (R_1', R_2') to this equation. Extending these results one can prove that for solutions (R_1', R_2') and (R_1'', R_2'') of this equation also $(R_1' \cup R_1'', R_2' \cap R_2'')$ is a solution. And extending Facts 3 and 4 one can prove that an extremal solution of a system of such equations (3.32) is determined – in the case of its existence – in its "first coordinate" as the intersection of the "first coordinates" of the extremal solutions of the single equations, and in its "second coordinate" as the union of the "second coordinates" of the extremal solutions of the single equations.

What we have achieved in this chapter is to give characterizations of the solvability of some simple types of fuzzy equations and systems of such equations whose usefulness in applications is already established. Using the solvability degrees of a fuzzy equation or of a system of such equations a characterization became possible to which degree – measured by the (pseudo-) metric ϱ_t connected with $\equiv_t$ – the fuzzy equation resp. the system of fuzzy equations has a solution. At the same time, this solvability degree gives information about best possible approximate solutions if there does not exist any real solution.

For single fuzzy equations this solvability degree can be calculated from the results proved in this chapter. For systems of fuzzy equations at least bounds for this solvability degree may be calculated: a lower bound was explicitely given in theorem 3.4, but from the fact that this lower bound also is an upper bound for the "n-th power" of the solvability degree it follows that each truth value whose "n-th power" is not smaller than this lower bound itself is an upper bound of the solvability degree.

Further research should test the usefulness of these solvability degrees for practical applications of fuzziness. Some discussions relating to this point will be given later on in chapter 5.

Chapter 4

Fuzzy controllers

4.1 The construction of fuzzy controllers

In very general terms, a fuzzy controller is some device that is intended to modelize some roughly known or roughly described process. Used together with the real process in feedback mode it may really act as a control device, and used in feedforward mode it can be used for predicting the process behaviour. In a certain sense thus a fuzzy controller realizes some vaguely known, vaguely described, or intrinsically rough "algorithm".[1]

The usual textbook examples of such vague "algorithms" comprise things like: cooking recipes; instructions for driving a car; rules governing the behaviour of a human operator who controls some chemical process; medical diagnosis etc.

All these examples and consequently the fuzzy controller too are supposed to be *rule-based*, i.e. constituted by a (finite) set of *control rules*. That is an idea the fuzzy controller has in common with a widespread naive, as well as formalized, understanding of algorithmic methods, but more general also with most methods of knowledge transfer and knowledge representation. The background idea is simply that the essential structure of such knowledge can be split into small portions of the type: **if** there is some situation **then** something is to be expected or has to be done.

That means that all those rule-based kinds of knowledge representation

[1]Of course here we refer to an intuitive, everyday understanding of the general notion of "algorithm"; it is not the formalized, mathematical notion of algorithm that should be the background idea. But, surely, there are (loose) connections between the formal and the present informal notion of algorithm.

refer to input as well as to output information. Usually the concrete input and output informations for each single case are seen as specific values of input or output variables. The sets of possible values of each such (input or output) variable are supposed to be determined in advance and very often considered as possible numerical values with respect to some scaling. But in general they are simply abstract sets of objects.

This last-mentioned fact is the background for a nice theoretical simplification. In any case, namely, one can suppose to have only one input and one output variable. For, if one has initially for example n input variables $\alpha_1, \dots, \alpha_n$ with respective universes $\mathcal{V}_1, \dots, \mathcal{V}_n$ of possible values, one can introduce a new variable $\bar{\alpha}$, often denoted $\bar{\alpha} = (\alpha_1, \dots, \alpha_n)$, whose set of possible values is the cartesian product $\mathcal{V}_1 \times \cdots \times \mathcal{V}_n$, i.e. whose values are n-tuples of values of $\alpha_1, \dots, \alpha_n$. And if one has m output variables $\beta_1, \dots, \beta_m$, one either considers one new, more complex output variable $\bar{\beta} = (\beta_1, \dots, \beta_m)$ or – normally much better – one takes into account m separate one-output fuzzy controllers which work parallel. Therefore in the following it is always enough to discuss the case of one input and one output variable.

The choice of values for the input and output variables is one of the topics with a very specialized understanding in the case of fuzzy controllers: all these input and output variables are taken as ***linguistic variables***,[2] i.e. their values are fuzzy subsets of suitable universes of discourse. Some specific interpretations are tied with this choice of "linguistic values". The first one, which was essential for naming these variables "linguistic" ones, is that it is supposed that these linguistic values are represented by common words from everyday language (like: ***high*** speed, ***hot*** water, ***very heavy*** rain etc.). These words are understood as meaning suitable fuzzy sets. The second one of these specific interpretations is that these fuzzy sets have to be fuzzy subsets of some suitable universe of discourse – and that the members of this suitable universe of discourse are considered as possible "true" values of some crisp, i.e. traditional variable loosely connected with the linguistic variable (like: speed, temperature of water, amount of rain per area and time etc.). And, finally, as a third and very essential interpretation one assumes that the "linguistic" values of the linguistic variable code unsharp, vague information about the ***possible "true" value*** of the traditional variable tied with the linguistic one.

Finally, crucial and constitutive for the idea of a fuzzy controller is that

[2]The notion of linguistic variable was introduced by ZADEH (1975); cf. also ZIMMERMANN (1985), BANDEMER/GOTTWALD (1989).

the control rules (and hence the fuzzy controller) shall also become applicable in situations, i.e. for values of the input variable which are not explicitly used in the (premises of the) control rules.

A fuzzy controller thus constitutes a connection, a relation between the values of the input and the output variables. Formally, hence, one intends to construct a fuzzy controller as a fuzzy relation (in the precise sense of section 2.2), even to identify it with such a fuzzy relation, and to get this relation out of the system of control rules.[3]

There is no uniform way to construct fuzzy controllers. But at present three strongly connected methods are used in most cases. To discuss them in more detail the input variable now will be denoted α, the output variable β. Both are linguistic variables; their universes of discourse (for the "linguistic" values) shall be $\mathcal{X}$ for α and $\mathcal{Y}$ for β.

The starting point in each case is a finite list of *control rules*

$$\textsf{if}\ \ \text{value}(\alpha) = A_i \quad \textsf{then}\ \ \text{value}(\beta) = B_i, \quad i = 1, \ldots, n, \tag{4.1}$$

the *generating family* that constitutes the fuzzy controller.

Because we suppose to have one fixed input and one fixed output variable we can follow common usage to take

$$A_i \models\!\!\Rightarrow B_i, \qquad i = 1, \ldots, n \tag{4.2}$$

as a shorthand notation for the generating family of control rules (4.1).

The pioneering approaches to fuzzy (logic) control given in MAMDANI (1974, 1976) and MAMDANI/ASSILIAN (1975) on the theoretical side and in HOLMBLAD/ØSTERGAARD (1982) with the first industrial application seemingly followed quite different strategies. Both started from a finite list (4.1) of control rules; but then MAMDANI wrote down almost directly a fuzzy relation – and HOLMBLAD/ØSTERGAARD used the idea that a specific input value of the input variable[4] would *activate* each control rule to a certain degree, and then this control rule would "act to that degree", i.e., contribute with its specific output to the total output according to this degree of activation.

[3]The terminology is not really unique up to now. Usually a fuzzy controller is considered as a fuzzy relation. In this sense a system of control rules isn't yet a controller, but *constitutes* one. And a (hardware or software) device actually having such a fuzzy relation implemented and thus realizing it then is a – practical – *realization* of a fuzzy controller.

[4]HOLMBLAD/ØSTERGAARD (1982) really did consider three independent input variables; but that is inessential here according to our foregoing remarks.

First, let us have a closer look at this method of activation degrees. Given the ***degree of activation*** $\operatorname{act}_j(A)$ of the j-th control rule $A_j \models\!\Rightarrow B_j$ with respect to the input value A, the output B_j of that j-th rule is "weighted", i.e. suitably modified into a new fuzzy set B_j^w depending on the degree $\operatorname{act}_j(A)$. At the end all the modified, weighted outputs "together", i.e. suitably superposed yield the final, total output B for the given input A.

Obviously, there are a lot of decisions to be made about the formal, mathematical character of the operations vaguely mentioned in this description of the approach through degrees of activation. The essential ones concern:

- the definition of the degree $\operatorname{act}_j(A)$ of activation of the j-th control rule for the input A;
- the definition of the "weighted" output B_j^w of the j-th rule if activated to degree $\operatorname{act}_j(A)$;
- the method of superposition of the weighted outputs B_j^w of all the (activated) control rules.

The degree of activation $\operatorname{act}_j(A)$ is, from the intuitive understanding of this process of activation, measuring the coincidence of the actual input A and the "rule input" A_j. If under the possibilistic reading both fuzzy sets A and A_j are taken as possibility distributions for the "true" input value it is intuitively appealing to take

$$\operatorname{act}_j(A) =_{def} \operatorname{hgt}(A \cap_{\boldsymbol{t}} A_j) \tag{4.3}$$

with respect to a suitable t-norm $\boldsymbol{t}$. HOLMBLAD/ØSTERGAARD (1982) take here $\boldsymbol{t} = \min$, such that they have

$$\operatorname{act}_j(A) = \sup_{x \in \mathcal{X}} \min\{\mu_A(x), \mu_{A_j}(x)\}. \tag{4.4}$$

The special degree of activation (4.3) can be understood as the degree of possibility to which control rule j may be the "correct" one. In some sense, thus, this is an *optimistic* view concerning the activation of control rule j by input A. A more *pessimistic* view may be represented for example by choosing

$$\operatorname{act}_j^*(A) =_{def} [A \equiv_{\boldsymbol{t}} A_j] \tag{4.5}$$

or some other "degree of equality" of A and A_j. This idea, however, seems to be *very* pessimistic because (4.5) may be read as meaning that control rule

j is activated (to a corresponding degree) just if the fuzzy input corresponds *exactly* to the situation covered by this j-th control rule. It is more suitable, perhaps, to approach the matter from the point of view that control rule j should become activated (to a corresponding degree) if the fuzzy input corresponds to a situation that is *covered* by the j-th rule. Then instead of (4.5) one would define

$$\mathrm{act}_j^\star(A) = [\![A \subseteq_t A_j]\!] = [\![\forall x(x \,\varepsilon\, A \to_t x \,\varepsilon\, A_j)]\!]. \tag{4.6}$$

But these "pessimistic" points of view have not till now been used for any realized fuzzy controllers. (Also, the theoretical consequences of those "pessimistic" approaches need further investigations.)

Accepting the possibilistic point of view a point $y \in \mathcal{Y}$ of the universe of discourse of the output variable β is the possible "true" output value if it is the "true" output of the j-th control rule and if that rule is activated. This idea, written down in our language for fuzzy sets, means for the "weighted" outputs B_j^w:

$$[\![y \,\varepsilon\, B_j^w]\!] = [\![y \,\varepsilon\, B_j \wedge_t \mathrm{act}_j(A)]\!]. \tag{4.7}$$

Here HOLMBLAD/ØSTERGAARD (1982) take $t = t_P$ = product and thus have in the usual notation with membership functions:

$$\mu_{B_j^w}(y) = \mathrm{act}_j(A) \cdot \mu_{B_j}(y) \quad \text{for all } y \in \mathcal{Y}. \tag{4.8}$$

Of course, other t-norms can do the job as well. Sometimes t = min is taken instead of (4.8).[5]

But it seems that for the modification of the output fuzzy sets the choice (4.8), i.e. $t = t_P$ in (4.7), is intuitively often more appealing then the choice t = min. For choosing t = min causes a modification of B_j that can be visualized as a "*cutting at level* $\mathrm{act}_j(A)$" in the sense that for all $x_0 \in \mathcal{X}$ with $\mu_{B_j}(x_0) > \mathrm{act}_j(A)$ one obtains through this modification $\mu_{B_j^w}(x_0) = \mathrm{act}_j(A)$ and for all other $x \in \mathcal{X}$ their membership degrees with respect to B_j and B_j^w coincide. With $t = t_P$ instead the membership function of B_j^w has a kind of similarity with that of B_j in the sense of some "proportional reduction" of all membership degrees.

As we mentioned in connection with (4.3) and (4.5), (4.6), for the definition of the degrees of activation, as well as for the definition of the weighted

[5] In the current literature as e.g. ZIMMERMANN (1985) one then often speaks of the max-dot resp. max-min inference method; the term "max" in these names refers to the method of superposition we discuss below – cf. (4.9) and the remarks that follow there.

outputs B_j^w, there exist conflicting, dual points of view connected with either the "optimistic" or the "pessimistic" perspective. Further variants for defining the weighted outputs will only be discussed later on, cf. (4.15).

The final point in this method of activation degrees is to make a superposition of the weighted outputs of the single (activated) control rules.[6] From the "optimistic" perspective of possibilistic understanding it is natural to collect, i.e. to sum up all the "local" control information provided by the outputs of the single control rules according to their degrees of activation, i.e. to take the final (fuzzy) output $B \in \mathbb{F}(\mathcal{Y})$ as

$$B =_{def} \bigcup_{i=1}^{n} B_i^w. \tag{4.9}$$

Together with formulas (4.3) and (4.7) that gives for each $y \in \mathcal{Y}$:[7]

$$\models y \varepsilon B \leftrightarrow \bigvee_{i=1}^{n} (y \varepsilon B_i \wedge_{t_1} \exists x(x \varepsilon A \wedge_t x \varepsilon A_i)) \tag{4.10}$$

what in the case of $\mathsf{LSC}(t_1)$ can be written as

$$\models y \varepsilon B \leftrightarrow \exists x \bigvee_{i=1}^{n} ((x \varepsilon A \wedge_t x \varepsilon A_i) \wedge_{t_1} y \varepsilon B_i) \tag{4.11}$$

with $\bigvee_{i=1}^{n}$ as the finite iteration of the max-disjunction.

Assuming furthermore $t_1 = t$ together with $\mathsf{LSC}(t)$ formula (4.11) can be transformed into the simpler form

$$\begin{aligned} \models y \varepsilon B \quad &\leftrightarrow \quad \exists x \bigvee_{i=1}^{n} (x \varepsilon A \wedge_t x \varepsilon A_i \wedge_t y \varepsilon B_i) \\ &\leftrightarrow \quad \exists x (x \varepsilon A \wedge_t \bigvee_{i=1}^{n} (x \varepsilon A_i \wedge_t y \varepsilon B_i)) \\ &\leftrightarrow \quad \exists x (x \varepsilon A \wedge_t (x, y) \varepsilon \bigcup_{i=1}^{n} (A_i \times_t B_i)). \end{aligned} \tag{4.12}$$

---

[6]It is inessential here to restrict the considerations to the activated rules only. If one takes into account all the rules, i.e. also those rules with degree of activation = 0, then because of (4.7) those rules contribute the empty fuzzy set $\emptyset$ as weighted output to that superposition (4.9) – and thus in any case do not influence the actual output at all. And precisely that last point will be of leading importance later on when discussing other variants of weighted outputs for the "pessimistic" case.

[7]Here as already in section 1.3 we assume that the biimplication $\leftrightarrow$ is based on some lower semicontinuous t-norm $\mathbf{t}'$: it is not necessary to be more specific because $\models H_1 \leftrightarrow H_2$ is equivalent to $[\![H_1]\!] = [\![H_2]\!]$ independent of the specific choice of $\mathbf{t}'$.

With the set theoretic notion of the full image $R''A$ of a fuzzy set A under a fuzzy relation R, cf. definition 2.9, formula (4.12) means

$$B = (\bigcup_{i=1}^{n}(A_i \times_{\mathbf{t}} B_i))''A. \tag{4.13}$$

On the more pessimistic side one tends to take instead of (4.9) for the superposition of the single weighted outputs B_j^w the idea that the final control output of the set of control rules should be chosen in such a way that it is proposed by all relevant, i.e. activated rules. This idea, hence, corresponds to taking the fuzzy output as

$$B =_{def} \bigcap_{i=1}^{n} B_i^w. \tag{4.14}$$

But now, with (4.7) any control rule which is not activated at all, i.e. for which one has $\text{act}_j(A) = 0$ as degree of activation, would contribute $B_j^w = \emptyset$ to the intersection (4.14) and hence cause the fuzzy output to become $B = \emptyset$. That, obviously, is counterintuitive. The idea, thus, leading to (4.14) has to be combined with another choice of the weighting process too, i.e. with another definition of B_j^w instead of (4.7). One possibility is to take a new weighted output $B_j^\star$ such that

$$[\![y \,\varepsilon\, B_j^\star]\!] =_{def} [\![\text{act}_j^\star(A) \rightarrow_{\mathbf{t}_1} y \,\varepsilon\, B_j]\!] \tag{4.15}$$

which needs the assumption $\mathsf{LSC}(\mathbf{t}_1)$. That approach, together with (4.14) and now with $\bigwedge_{i=1}^{n}$ as the finite iteration of the min-conjunction, gives

$$\models y \,\varepsilon\, B \leftrightarrow \bigwedge_{i=1}^{n}(\text{act}_i^\star(A) \rightarrow_{\mathbf{t}_1} y \,\varepsilon\, B_i). \tag{4.16}$$

With a degree of activation according to (4.3) then one has

$$\begin{aligned}
\models y \,\varepsilon\, B \quad &\leftrightarrow \quad \bigwedge_{i=1}^{n}(\exists x(x \,\varepsilon\, A \wedge_{\mathbf{t}} x \,\varepsilon\, A_i) \rightarrow_{\mathbf{t}_1} y \,\varepsilon\, B_i) \\
&\leftrightarrow \quad \bigwedge_{i=1}^{n} \forall x(x \,\varepsilon\, A \wedge_{\mathbf{t}} x \,\varepsilon\, A_i \rightarrow_{\mathbf{t}_1} y \,\varepsilon\, B_i) \qquad (4.17) \\
&\leftrightarrow \quad \forall x \bigwedge_{i=1}^{n}(x \,\varepsilon\, A \wedge_{\mathbf{t}} x \,\varepsilon\, A_i \rightarrow_{\mathbf{t}_1} y \,\varepsilon\, B_i) \qquad (4.18)
\end{aligned}$$

using proposition 1.27 (ii). Assuming furthermore $t_1 = t$ together with $\mathsf{LSC}(t)$ yields by proposition 1.19 (i)

$$\begin{aligned} \models y \varepsilon B \quad &\leftrightarrow \quad \forall x \bigwedge_{i=1}^{n} (x \varepsilon A \rightarrow_t (x \varepsilon A_i \rightarrow_t y \varepsilon B_i)) \\ &\leftrightarrow \quad \forall x (x \varepsilon A \rightarrow_t \bigwedge_{i=1}^{n} (x \varepsilon A_i \rightarrow_{t_1} y \varepsilon B_i)). \end{aligned} \tag{4.19}$$

With the degree of activation (4.6) instead one has

$$\begin{aligned} \models y \varepsilon B \quad &\leftrightarrow \quad \bigwedge_{i=1}^{n} (\forall x (x \varepsilon A \rightarrow_t x \varepsilon A_i) \rightarrow_{t_1} y \varepsilon B_i) \\ &\leftrightarrow \quad \bigwedge_{i=1}^{n} \exists x ((x \varepsilon A \rightarrow_t x \varepsilon A_i) \rightarrow_{t_1} y \varepsilon B_i) \end{aligned} \tag{4.20}$$

which, even for $t = t_1$ seem to be hard to transform into a much simpler and more appealing form like (4.19).

MAMDANI's way of approach toward fuzzy control in MAMDANI (1974, 1976) immediately aimed at the transformation of each control rule $A_i \models\!\!\Rightarrow B_i$ into a fuzzy relation $\Upsilon(A_i \models\!\!\Rightarrow B_i)$ in $\mathcal{X} \times \mathcal{Y}$ *coding* that rule and at the "collection" of all those codes of the single control rules into a fuzzy relation R, the fuzzy controller. He used as the final input-output-behaviour of the fuzzy controller R the approach of taking

$$B := R''A \tag{4.21}$$

as the output of the fuzzy controller R for the input A, thus using an earlier proposal by ZADEH (1973).[8]

Writing as usual R_i for the code of control rule i, i.e. taking

$$R_i =_{def} \Upsilon(A_i \models\!\!\Rightarrow B_i), \tag{4.22}$$

MAMDANI's way of "collecting" all these coding relations R_i into one fuzzy relation R was to "sum them up" in the sense of taking their union (cf. definition 2.5) and thus having

$$R = \bigcup_{i=1}^{n} \Upsilon(A_i \models\!\!\Rightarrow B_i) = \bigcup_{i=1}^{n} R_i \tag{4.23}$$

[8] ZADEH (1973) did not use the name "full image of A under R" for (4.21) but spoke of the "compositional rule of inference" and wrote $A \circ R$ for our $R''A$.

independent of the different ways $R_i = \Upsilon(A_i \models\!\!\Rightarrow B_i)$ may have been chosen.

There is no general rule governing the coding procedure for the control rules. But there is a kind of agreement that ideally these codes R_i should have the property that one has to have

$$R_i'' A_i = B_i \tag{4.24}$$

for each $i = 1, \ldots, n$. Furthermore, there is often the background idea that the relation R_i has to reflect in some sense the fact that in its original, full-length form (4.1) the control rules are seen as fuzzy implications. Thus one often constructs $R_i = \Upsilon(A_i \models\!\!\Rightarrow B_i)$ in a manner which is analogous to well known possibilities one has in two-valued classical propositional logic to express the implication connective using for example the conjunction, negation, and disjunction connectives. Of course, there are many such possibilities if one does not restrict oneself to the shortest and simplest ones.

Prefered candidates for ways to build the relations R_i are often:

$$\Upsilon(A_i \models\!\!\Rightarrow B_i) = A \times B, \tag{4.25}$$

$$\Upsilon(A_i \models\!\!\Rightarrow B_i) = (A \times B) \cup (\overline{A} \times Y), \tag{4.26}$$

$$\Upsilon(A_i \models\!\!\Rightarrow B_i) = (\overline{A} \times Y) \cup (X \times B). \tag{4.27}$$

For many more such proposals cf. MIZUMOTO (1982), MIZUMOTO/ZIMMERMANN (1982).

MAMDANI originally had chosen version (4.25) thereby meeting condition (4.24) for each control rule.

If one compares (4.23) with (4.13) one immediately recognizes that there is a very close connection between the approaches by HOLMBLAD/ØSTERGAARD (1982) and by MAMDANI (1976), notwithstanding the fact that both these approaches quite naturally generalize into different directions.

The third method of approach toward fuzzy controllers, besides those by HOLMBLAD/ØSTERGAARD and by MAMDANI, also intends to consider a fuzzy controller as a fuzzy relation R which is constituted by a set (4.1) of control rules. And that approach too accepts the input-output-behaviour (4.21) in the sense of the "compositional rule of inference" .

As a side remark let us mention that, of course, one is not forced to accept that approach through (4.21) and definition 2.9, i.e. to accept the "compositional rule of inference" as the decisive way to convert an input into an output. But at present there does not exist any other reasonable proposal here – besides, in some sense, the direct approach of HOLMBLAD/ØSTERGAARD which, however, in special cases like (4.13) reduces to this method too.

The essential idea now is that this type of input-output-behaviour of the fuzzy controller R should prove to be a generalization of the input-output-relation (4.24) immmediately pretended by the control rules, i.e. that from input A_i the output B_i should be produced by R, i.e.

$$R''A_i = B_i, \qquad \text{for all } i = 1, \ldots, n \tag{4.28}$$

has to be the case.

In the same way as (4.24), for each $i = 1, \ldots, n$, can be understood as a condition which the fuzzy relation R has to meet for the fuzzy sets A_i, B_i given by the i-th control rule, the condition (4.28) can be seen as constitutive for the fuzzy controller R. In other words, as (4.24), for each $i = 1, \ldots, n$, can be considered as an equation which has to be solved with respect to an unknown fuzzy relation R_i, the condition (4.28) can be considered as a system of equations[9] which has to be solved with respect to an unknown fuzzy relation R.

This third kind of approach has stimulated many studies concerning solutions and sets of solutions of such relation equations, as well as of systems of them, cf. DINOLA/SESSA/PEDRYCZ/SANCHEZ (1989), and is also the background for further discussion in this book.

4.2 The problem of interaction

The second one (4.23) of these ways to construct a fuzzy controller out of the fuzzy relations $\Upsilon(A_i \models\!\!\Rightarrow B_i)$ which code its single control rules, sometimes is not in accordance with (4.28), i.e. with the idea that (4.24) ought also to hold true for each control rule in the case that instead of R_i the fuzzy controller R is taken into account. This possibility was seen quite early among control engineers, cf. e.g. CZOGALA/PEDRYCZ (1981). That means that it may happen that there finally is a control rule $A_k \models\!\!\Rightarrow B_k$ such that

$$B_k \neq R''A_k \tag{4.29}$$

is the case. But that means that the final fuzzy controller R from (4.23) sometimes may act in a different way than was intended in the set of control rules and the coding procedure. As we will say for this situation: the control rules may interact.

[9] Because of the specific form of these equations they usually are denoted as *(fuzzy) relation(al) equations.*

4.2.1 General results on interactivity

Definition 4.1 *A generating family of control rules $A_i \models\!\!\Rightarrow B_i$, $i \in I = \{1,\dots,n\}$, for a fuzzy controller R is called* interactive *iff there exists some index $k \in I$ such that $B_k \neq R''A_k$; otherwise the generating family of control rules is called* non-interactive.

As in this definition we will, in the present section, always use the notation $I = \{1,\dots,n\}$ for simplicity of notation.

Because, according to definition 2.3, the condition of equality for fuzzy sets can be split into two conditions of being subsets of one another, the problem of non-interactivity of the families of generating rules can be split too.

Definition 4.2 *Suppose that R is a fuzzy controller to realize the family of control rules $A_i \models\!\!\Rightarrow B_i$. We say that R has the* superset property *w. r. t. the generating family (4.2) iff*

$$R''A_i \supset B_i \quad \text{for all } i \in I\ , \tag{4.30}$$

and we say that R has the subset property *w. r. t. this family iff*

$$R''A_i \subset B_i \quad \text{for all } i \in I \tag{4.31}$$

hold true.

We always assume in the following considerations that the fuzzy controller R was constructed from a family (4.1) of control rules.

Proposition 4.1 (*i*) *A sufficient condition that the superset property holds true for this fuzzy controller R is that for all $i \in I$ and all $y \in \mathcal{Y}$ there exists some $k \in I$ and some $x_0 \in \mathcal{X}$ such that*

$$[\![y \,\varepsilon\, B_i]\!] \leq [\![(x_0 \,\varepsilon\, A_i \wedge_t ((x_0, y) \,\varepsilon\, \Upsilon(A_k \models\!\!\Rightarrow B_k)))]\!].$$

If $\mathcal{X}$ is a finite set, this condition is necessary too.

(*ii*) *A necessary and sufficient condition that the subset property holds true for this fuzzy controller R is that for all $i, j \in I$, all $x \in \mathcal{X}$ and all $y \in \mathcal{Y}$ one has*

$$[\![y \,\varepsilon\, B_i]\!] \geq [\![x \,\varepsilon\, A_i \wedge_t ((x, y) \,\varepsilon\, \Upsilon(A_j \models\!\!\Rightarrow B_j))]\!].$$

Proof. By the corresponding definitions we immediately have that for all $y \in \mathcal{Y}$:

$$\begin{aligned}
&[\![y \varepsilon R''A]\!] \\
&\quad = [\![\exists x(x \varepsilon A \wedge_t \exists_I k((x,y) \varepsilon \Upsilon(A_k \models\!\!\Rightarrow B_k)))]\!] \\
&\quad = [\![\exists x \exists_I k(x \varepsilon A \wedge_t ((x,y) \varepsilon \Upsilon(A_k \models\!\!\Rightarrow B_k)))]\!].
\end{aligned}$$

Therefore the superset property means that for all $i \in I$ and $y \in \mathcal{Y}$

$$[\![y \varepsilon B_i]\!] \leq [\![\exists x \exists_I j(x \varepsilon A_i \wedge_t ((x,y) \varepsilon \Upsilon(A_j \models\!\!\Rightarrow B_j)))]\!]$$

has to be the case; and the subset property means that for all $i \in I$ and $y \in \mathcal{Y}$:

$$[\![y \varepsilon B_i]\!] \geq [\![\exists x \exists_I k(x \varepsilon A_i \wedge_t ((x,y) \varepsilon \Upsilon(A_j \models\!\!\Rightarrow B_j)))]\!]$$

has to be the case.

Because $\exists$ means taking the supremum of the corresponding truth degrees, and the generating families of control rules are always supposed to be finite, we have that the superset property means that for all $i \in I$ and all $y \in \mathcal{Y}$ there exists some $k \in I$ such that

$$\begin{aligned}
[\![y \varepsilon B_i]\!] &\leq [\![\exists x(x \varepsilon A_i \wedge_t ((x,y) \varepsilon \Upsilon(A_k \models\!\!\Rightarrow B_k)))]\!] \\
&= \sup_{x \in \mathcal{X}} [\![x \varepsilon A_i \wedge_t ((x,y) \varepsilon \Upsilon(A_k \models\!\!\Rightarrow B_k))]\!]
\end{aligned} \tag{4.32}$$

has to be the case, and correspondingly the subset property means that for all $i, j \in I$ and all $y \in \mathcal{Y}$

$$\begin{aligned}
[\![y \varepsilon B_i]\!] &\geq [\![\exists x(x \varepsilon A_i \wedge_t ((x,y) \varepsilon \Upsilon(A_j \models\!\!\Rightarrow B_j)))]\!] \\
&= \sup_{x \in \mathcal{X}} [\![x \varepsilon A_i \wedge_t ((x,y) \varepsilon \Upsilon(A_j \models\!\!\Rightarrow B_j))]\!]
\end{aligned} \tag{4.33}$$

has to be the case. Now it is obvious that the condition given in (ii) is necessary and sufficient for the subset property to hold true.

Of course the existence of some $x_0 \in \mathcal{X}$ such that

$$[\![y \varepsilon B_i]\!] \leq [\![(x_0 \varepsilon A_i \wedge_t ((x_0,y) \varepsilon \Upsilon(A_k \models\!\!\Rightarrow B_k)))]\!] \tag{4.34}$$

is the case is a sufficient condition for (4.32) to hold true. The crucial point for the difficulties with necessary conditions for (4.32) is that there the supremum is on the right hand side of the $\leq$-condition. But for a finite universe of discourse $\mathcal{X}$ the supremum becomes a maximum and the difficulty disappears. Hence (i) is proved too. QED

Unfortunately, the conditions we obtained in this proposition are not very handy ones. Therefore we will look for simpler necessary or sufficient conditions for the superset, as well as the subset condition. Without restriction to special cases how to choose the "local" fuzzy relations $R_i = \Upsilon(A_i \models\!\!\!\Rightarrow B_i)$ one has the following corollary.

Corollary 4.2 *(i) A necessary condition for the superset property of R is*

$$\bigwedge_{i\in I}(\mathrm{hgt}\,(B_i) \le \mathrm{hgt}\,(A_i)). \tag{4.35}$$

(ii) A sufficient condition for the superset property of R is

$$\bigwedge_{i\in I}(B_i \subset \Upsilon(A_i \models\!\!\!\Rightarrow B_i)''A_i). \tag{4.36}$$

(iii) A necessary and sufficient condition for the subset property of R is

$$\bigwedge_{i,j\in I}(\Upsilon(A_i \models\!\!\!\Rightarrow B_i)''A_j \subset B_j). \tag{4.37}$$

Sometimes it is useful to have not only absolute conditions concerning the superset and the subset property, but also relative ones. For that purpose we give the following definition.

Definition 4.3 *Consider a control rule $A_i \models\!\!\!\Rightarrow B_i$ and two of its realizations R_i^1, R_i^2 as a fuzzy relation. Then we call R_i^1 a* stronger realization *(of that rule) as R_i^2 and R_i^2 a* weaker realization *(of that rule) as R_i^1 iff $R_i^1 \subset R_i^2$ holds true. In the same manner, if R^1 and R^2 are fuzzy relations which realize a given set (4.1) of control rules for a fuzzy controller, we call R^1 a* stronger *realization (of (4.1)) as R^2 or R^2 a* weaker *realization (of (4.1)) as R^1 iff $R^1 \subset R^2$.*

Corollary 4.3 *Let R^1, R^2 be two realizations of a set of control rules (4.1) as a fuzzy relation.*

(i) If R^1 is a stronger realization as R^2 and R^1 has the superset property, then R^2 has the superset property too.

(ii) If R^1 is a weaker realization as R^2 and R^1 has the subset property, then R^2 has the subset property too.

Proof. Both claims are immediate consequences of the monotonicity, proposition 2.18(ii), of the full image of a fuzzy set under a fuzzy relation with respect to the inclusion of the relations involved and of the two foregoing definitions. QED

Of course, if the coding relation R of a fuzzy controller which is constituted by a set (4.1) of control rules is built up in MAMDANI's way (4.9), (4.25) out of the coding relations $R_i = A_i \times B_i$ of the single rules of (4.1), and if one has two coding procedures for the single rules such that each relation R_i^1 produced by the first coding procedure is a weaker version than the fuzzy relation R_i^2 produced by the second coding procedure, then R^1 is a weaker version than R^2. But what is essential here is not the specific way in which one "collects" the individual information R_i to get the full coding relation R for the fuzzy controller, but only a certain monotonicity property again.

Proposition 4.4 *Suppose there are two procedures to obtain for the control rules $A_i \models\!\!\Rightarrow B_i$ of a fuzzy controller with generating family (4.1) their realizations R_i^1, R_i^2 as fuzzy relations. Suppose additionally that the method by which the complete coding relation R for a fuzzy controller is constructed out of the local relations R_i is monotonuous with respect to inclusion. Then R^1 is a stronger realization of a set of control rules (4.2) if for each of the rules $A_i \models\!\!\Rightarrow B_i$ the coding fuzzy relation R_i^1 is a stronger realization of that control rule than the coding relation R_i^2.*

Proof. If R_i^1 is always a stronger realization than R_i^2, by definition 4.3 that means that $R_i^1 \subset R_i^2$ is always the case. Therfore, by the monotonic dependence of each R^k of its constituents R_i^k, $k = 1, 2$ one immediately has that $R^1 \subset R^2$. QED

The essential property of monotonic dependence of the coding relation R of a fuzzy controller from its local constituents, the coding relations R_i of the single control rules, is not only fulfilled for MAMDANI's method of taking for R the union of all the $R_i = A_i \times B_i$, but for example also for the method (4.14) to take as R the intersection of suitable R_i.

In the case of the coding procedures (4.25) to (4.27) for all fuzzy sets $A \in I\!F(\mathcal{X}), B \in I\!F(\mathcal{Y})$ we have the inclusions

$$A \times B \subset (A \times B) \cup (\overline{A} \times Y) \subset (\overline{A} \times Y) \cup (X \times B) \tag{4.38}$$

and therefore, if these coding procedures are used within the collecting strategy (4.9), coding procedure (4.25) is the strongest one and coding procedure (4.27) the weakest one of these three.

Thus, in (4.38) the superset property is transmitted "from left to right" and the subset property "from right to left".

To get more concrete results we now will consider as special cases those where the coding procedure for the single control rules is given by (4.25) to (4.27). Additionally, in these three cases we suppose that the codes of the single control rules are collected together via (4.23) to give the fuzzy relation R which codes the whole fuzzy controller. Remember for the following considerations that the inequalities which appeared in proposition (4.1) as necessary and sufficient conditions for the superset or subset property have on their right sides a reference to a conjunction connective, i.e. to a t-norm. This t-norm will now always be denoted $\boldsymbol{t}$ without an index. It has to be distinguished from further t-norms which will be considered and which come into consideration by the special coding procedures for the control rules. In order to be not too specific here we intend to allow that in principle the t-norm $\boldsymbol{t}$ and these other ones are different.

4.2.2 Coding rules by $\Upsilon(A \models\!\!\Rightarrow B) = A \times_{\boldsymbol{t}} B$

We assume that the coding procedure for the single control rules is simply to take the cartesian product of input and output of the rule. For that cartesian product it is supposed that it refers to a t-norm $\boldsymbol{t}_1$.

Proposition 4.5 *Sufficient conditions for R to have the superset property are:*

(*i*) *for each t-norm* $\boldsymbol{t}$:

$$\mathrm{hgt}(A_j) = 1 \quad \textit{for all } j \in I;$$

(*ii*) *if* $\boldsymbol{t} = \boldsymbol{t}_1 = \min$:

$$\mathrm{hgt}(B_j) \le \mathrm{hgt}(A_j) \quad \textit{for all } j \in I.$$

Proof. By (4.32) we have as a necessary and sufficient condition for the superset property in the present case that for all $i \in I$ and all $y \in \mathcal{Y}$ there exists some $k \in I$ such that

$$\begin{aligned} &[\![y \,\varepsilon\, B_i]\!] \\ &\quad \le \sup_{x \in \mathcal{X}} [\![(x \,\varepsilon\, A_i \wedge_{\boldsymbol{t}} ((x,y) \,\varepsilon\, (A_k \times_{\boldsymbol{t}_1} B_k)))]\!] \\ &\quad \le \sup_{x \in \mathcal{X}} [\![x \,\varepsilon\, A_i \wedge_{\boldsymbol{t}} (x \,\varepsilon\, A_k \wedge_{\boldsymbol{t}_1} y \,\varepsilon\, B_k)]\!]. \end{aligned} \tag{4.39}$$

Now choosing $k = i$ one gets as a sufficient condition

$$[\![y \varepsilon B_i]\!] \leq \sup_{x \in \mathcal{X}} [\![(x \varepsilon A_i \wedge_{\mathbf{t}} (x \varepsilon A_i \wedge_{\mathbf{t}_1} y \varepsilon B_i)]\!], \tag{4.40}$$

which because of $\mathrm{hgt}\,(A_i) = \sup_{x \in \mathcal{X}} [\![x \varepsilon A_i]\!]$ and the left continuity of $\mathbf{t}$ and of $\mathbf{t}_1$ can be rewritten as

$$[\![y \varepsilon B_i]\!] \leq \mathrm{hgt}\,(A_i)\,\mathbf{t}\,(\mathrm{hgt}\,(A_i)\,\mathbf{t}_1\,[\![y \varepsilon B_i]\!]). \tag{4.41}$$

Thus it is obvious that condition (i) is a sufficient one for (4.41) to hold true for all $i \in I$ and all $y \in \mathcal{Y}$, and hence also for (4.39) to hold true for all $i \in I$ and all $y \in \mathcal{Y}$.

Furthermore, if we assume that $\mathbf{t} = \mathbf{t}_1 = \min$ is the case, (4.41) can essentially be simplified and is equivalent to the fact that

$$\begin{aligned} [\![y \varepsilon B_i]\!] &\leq \min\{\mathrm{hgt}\,(A_i), \min\{\mathrm{hgt}\,(A_i), [\![y \varepsilon B_i]\!]\}\} \\ &\leq \min\{\mathrm{hgt}\,(A_i), [\![y \varepsilon B_i]\!]\} \end{aligned} \tag{4.42}$$

holds true for all $i \in I$ and all $y \in \mathcal{Y}$. But this means that (ii) is also a sufficient condition for R to have the superset property. QED

Perhaps at first sight condition (i) of this proposition appears strong, but it is quite reasonable in most applications of fuzzy controllers. The reason is simply that usually the typical input values A_i for the control rules characterize some kinds of "standard situations" for the "true" values of the input variable. Therefore these input fuzzy sets A_i are usually normal and quite often even unimodal. In the specific situation with the min-operator as the only t-norm under consideration, by condition (ii) an even weaker assumption already suffices, which additionally is easy to check and to handle. Hence, the superset property of a fuzzy controller is quite a weak condition which is easy to meet in applications.

Proposition 4.6 *Sufficient conditions for R to have the subset property are:*

(i) without any restriction concerning the t-norm involved:

$A_i \cap_{\mathbf{t}} A_j = \emptyset \quad$ *for all* $i, j \in I, i \neq j$;

(ii) if $u\,\mathbf{t}\,(v\,\mathbf{t}_1\,w) \leq (u\,\mathbf{t}\,v)\,\mathbf{t}_1\,w$ *for all* $u, v, w, \in [0, 1]$:

$\mathrm{hgt}(B_i)\,\mathbf{t}_1\,\mathrm{hgt}\,(A_i \cap_{\mathbf{t}} A_j) = 0 \quad$ *for all* $i \in I, i \neq j$;

(iii) if $\mathbf{t} = \mathbf{t}_D$ *is the drastic product*

$\mathrm{hgt}(A_i \cap A_j) < 1 \quad$ *for all* $i, j \in I, i \neq j$.

Proof. By (4.33) we have as a necessary and sufficient condition for the subset property in the present case that for all $i, j \in I$ and all $y \in \mathcal{Y}$

$$
\begin{aligned}
&[\![y \,\varepsilon\, B_i]\!] \\
&\quad\geq \sup_{x \in \mathcal{X}} [\![x \,\varepsilon\, A_i \wedge_{\boldsymbol{t}} ((x,y) \,\varepsilon\, (A_j \times_{\boldsymbol{t}_1} B_j))]\!] \\
&\quad\geq \sup_{x \in \mathcal{X}} [\![x \,\varepsilon\, A_i \wedge_{\boldsymbol{t}} (x \,\varepsilon\, A_j \wedge_{\boldsymbol{t}_1} y \,\varepsilon\, B_j)]\!]. \qquad (4.43)
\end{aligned}
$$

has to be the case. Using the left continuity of the t-norms involved here one gets as in the last proof as an equivalent condition to (4.43) that for all $i, j \in I$ and all $y \in \mathcal{Y}$

$$[\![y \,\varepsilon\, B_i]\!] \geq \operatorname{hgt}(A_i) \wedge_{\boldsymbol{t}} (\operatorname{hgt}(A_j) \wedge_{\boldsymbol{t}_1} [\![y \,\varepsilon\, B_j]\!]) \qquad (4.44)$$

has to be the case. In the case that $i = j$ inequality (4.44) obviously always holds true. But for $i \neq j$ the only reasonable way to get a sufficient condition for (4.44) and hence for (4.43) is to make sure that the right-hand side of inequality (4.44) is always zero. Because of

$$\operatorname{hgt}(A_i) \wedge_{\boldsymbol{t}} \operatorname{hgt}(A_j) \geq \operatorname{hgt}(A_i) \wedge_{\boldsymbol{t}} (\operatorname{hgt}(A_j) \wedge_{\boldsymbol{t}_1} [\![y \,\varepsilon\, B_j]\!])$$

for all $i, j \in I$ and all $y \in \mathcal{Y}$, the sufficiency of condition (i) immediately follows.

The inequalities

$$
\begin{aligned}
&(\operatorname{hgt}(A_i) \wedge_{\boldsymbol{t}} \operatorname{hgt}(A_j)) \wedge_{\boldsymbol{t}_1} \operatorname{hgt}(B_j) \\
&\quad\geq \operatorname{hgt}(A_i) \wedge_{\boldsymbol{t}} (\operatorname{hgt}(A_j) \wedge_{\boldsymbol{t}_1} \operatorname{hgt}(B_j)) \\
&\quad\geq \operatorname{hgt}(A_i) \wedge_{\boldsymbol{t}} (\operatorname{hgt}(A_j) \wedge_{\boldsymbol{t}_1} [\![y \,\varepsilon\, B_j]\!]),
\end{aligned}
$$

the first one of them a consequence of the specific assumption of (ii), together with (4.44) prove that (ii) is also a sufficient condition for the subset property of R.

Finally, condition (iii) is a sufficient one because of the equivalence:

$$A_i \cap_{\boldsymbol{t}} A_j = \emptyset \Leftrightarrow \operatorname{hgt}(A_i \cap A_j) < 1 \quad \text{for } \boldsymbol{t} = \boldsymbol{t}_D$$

for all fuzzy sets A_i, A_j, and of the fact that in the case of $A_i \cap_{\boldsymbol{t}} A_j = \emptyset$ one has

$$[\![x \,\varepsilon\, A_i \wedge_{\boldsymbol{t}} (x \,\varepsilon\, A_j \wedge_{\boldsymbol{t}_1} y \,\varepsilon\, B_j)]\!] = 0$$

independent of $[\![y \,\varepsilon\, B_j]\!]$. Here the left continuity is not used in the arguments.

QED

Condition (i) of this proposition is simply the pairwise disjointness of the family of input fuzzy sets A_i of the fuzzy controller under consideration. Caused by the possible choice of an interactive t-norm for $\boldsymbol{t}$ this does not necessarily mean the pairwise disjointness of the supports supp (A_i) of these input fuzzy sets, but nevertheless is a quite strong demand. Condition (iii) is essentially the same condition, only reformulated using specific properties of the drastic product. Condition (ii), finally, is some weakening of this pairwise disjointness guided by the height, i.e. the "amount of subnormality" of B_i, but does not look very manageable. Hence, the sufficient conditions this last proposition gives for the subset property are relatively strong ones. And the chosen way of proof indicates in addition that much simpler sufficient conditions for the subset property of R seem not – or at least not easily – available. That means the subset property for R is a much stronger demand than the superset property.

4.2.3 Coding rules by $\Upsilon(A \mapsto B) = (A \times_{\boldsymbol{t}_1} B) \cup_{\boldsymbol{t}_2} (\overline{A} \times_{\boldsymbol{t}_1} Y)$

We now assume that the coding procedure for the single control rules is more involved. The main idea, however, is near to case 1. The coding there of the control rules $(A_i \mapsto B_i)$ by the cartesian product $A_i \times_{\boldsymbol{t}_1} B_i$ can be seen as a kind of coding of a fuzzy implication

$$\textsf{if}\ \text{INPUT} = A_i\ \textsf{then}\ \text{OUTPUT} = B_i. \tag{4.45}$$

Now the i-th control rule is not read as precisely this fuzzy implication but instead as the fuzzy implication

$$\textsf{if}\ \text{INPUT} = A_i\ \textsf{then}\ \text{OUTPUT} = B_i\ \textsf{else}\ \text{OUTPUT} = \text{ANYTHING}$$

with the **else**-clause furthermore understood as saying

$$\textsf{if}\ \text{INPUT} = \textsf{not-}A_i\ \textsf{then}\ \text{OUTPUT} = \text{ANYTHING} \tag{4.46}$$

and being disjunctively appended to (4.45). Then the following conditions result.

Proposition 4.7 *Sufficient conditions for R to have the superset property are:*

(*i*) *in any case*

$\text{hgt}(A_j) = 1$ *for all* $j \in I$;

(ii) for each $i \in I$ there exists some $k \in I$ such that

$$A_i^{\geq 1} \cap (\mathcal{X} \setminus \mathrm{supp}(A_k)) \neq \emptyset;$$

(iii) if $\mathbf{t}_1 = \mathbf{t}_2 = \mathbf{t}_L$ or $\mathbf{t}_1 = \mathbf{t}_2 = \min$, then

$$\mathrm{hgt}(B_i) \leq \mathrm{hgt}\Big(A_i \cap_{\mathbf{t}} \bigcup_{j \in I} \overline{A_j}\Big) \quad \textit{for all } j \in I;$$

and assuming additionally that $\mathbf{t} = \min$ one also gets

(iv) if either $\mathbf{t} = \mathbf{t}_1 = \mathbf{t}_2 = \min$ or $\mathbf{t} = \min$ together with $\mathbf{t}_1 = \mathbf{t}_2 = \mathbf{t}_L$, then

$$\mathrm{hgt}(B_j) \leq \mathrm{hgt}(A_j) \quad \textit{for all } j \in I.$$

Proof. By (4.32) we have as a necessary and sufficient condition for the superset property in the present case that for all $i \in I$ and all $y \in \mathcal{Y}$ there exists some $k \in I$ such that

$$\begin{aligned} [\![y \,\varepsilon\, B_i]\!] &\leq [\![\exists x (x \,\varepsilon\, A_i \wedge_{\mathbf{t}} ((x,y) \,\varepsilon\, \Upsilon(A_k \Mapsto B_k)))]\!] \\ &= \sup_{x \in \mathcal{X}} [\![x \,\varepsilon\, A_i \wedge_{\mathbf{t}} ((x,y) \,\varepsilon\, (A_k \times_{\mathbf{t}_1} B_k) \cup_{\mathbf{t}_2} (\overline{A_k} \times_{\mathbf{t}_1} Y)]\!] \\ &= \sup_{x \in \mathcal{X}} [\![x \,\varepsilon\, A_i \wedge_{\mathbf{t}} ((x \,\varepsilon\, A_k \wedge_{\mathbf{t}_1} y \,\varepsilon\, B_k) \vee_{\mathbf{t}_2} x \,\varepsilon\, \overline{A_k})]\!]. \end{aligned} \tag{4.47}$$

Using $[\![x \,\varepsilon\, \overline{A_k}]\!] \geq 0$ together with the monotonicity of $\vee_{\mathbf{t}_2}$ and $\wedge_{\mathbf{t}}$ and with condition (S3) in the definition 1.1 of t-conorms, we get as a sufficient condition

$$[\![y \,\varepsilon\, B_i]\!] \leq \sup_{x \in \mathcal{X}} [\![x \,\varepsilon\, A_i \wedge_{\mathbf{t}} (x \,\varepsilon\, A_k \wedge_{\mathbf{t}_1} y \,\varepsilon\, B_k)]\!], \tag{4.48}$$

which is exactly condition (4.39). Thus (i) is also a sufficient condition in this case.

Furthermore, it is surely sufficient for (4.47) to have for each $i \in I$ and $y \in \mathcal{Y}$

$$\sup_{x \in \mathcal{X}} [\![x \,\varepsilon\, A_i \wedge_{\mathbf{t}} ((x \,\varepsilon\, A_k \wedge_{\mathbf{t}_1} y \,\varepsilon\, B_k) \vee_{\mathbf{t}_2} x \,\varepsilon\, \overline{A_k})]\!] = 1 \tag{4.49}$$

for a suitable $k \in I$. Therefore it is sufficient too to always find $x \in \mathcal{X}$ and $k \in I$ such that

$$[\![x \,\varepsilon\, A_i]\!] = [\![x \,\varepsilon\, \overline{A_k}]\!] = 1.$$

But this precisely is condition (ii).

Now suppose that $t_1 = t_2 = t_L$. Then one has that for all $u, v \in [0,1]$ the equality

$$(u \& v) \uplus \overline{u} = \overline{u} \vee v$$

holds true where $\overline{u} = 1 - u$ is used. Thus, (4.47) becomes the form: for all $i \in I$ and all $y \in \mathcal{Y}$ there exists some $k \in I$ such that

$$\begin{aligned} [\![y \,\varepsilon\, B_i]\!] &\leq \sup_{x \in \mathcal{X}} [\![x \,\varepsilon\, A_i \wedge_t (x \,\varepsilon\, \overline{A_k} \vee y \,\varepsilon\, B_k)]\!] \\ &= \sup_{x \in \mathcal{X}} [\![(x \,\varepsilon\, A_i \cap_t \overline{A_k}) \vee (x \,\varepsilon\, A_i \wedge_t y \,\varepsilon\, B_k)]\!] \\ &= \max\{ \sup_{x \in \mathcal{X}} [\![x \,\varepsilon\, A_i \cap_t \overline{A_k}]\!], \sup_{x \in \mathcal{X}} [\![x \,\varepsilon\, A_i \wedge_t y \,\varepsilon\, B_k]\!] \}. \end{aligned}$$

By the left continuity of t this last form can be written as

$$[\![y \,\varepsilon\, B_i]\!] \leq \max\{ \operatorname{hgt}(A_i \cap_t \overline{A_k}), \operatorname{hgt}(A_i)\, t\, [\![y \,\varepsilon\, B_k]\!] \} \tag{4.50}$$

and immediately proves the sufficiency of the condition that for each $i \in I$ there exists some $k \in I$ such that

$$\operatorname{hgt}(B_i) \leq \operatorname{hgt}(A_i \cap_t \overline{A_k}).$$

But now it is a routine matter to verify the sufficiency of condition (iii) for the present case.

Assuming furthermore that $t = \min$, (4.50) becomes the equivalent form: for all $i \in I$ and all $y \in \mathcal{Y}$ there exists some $k \in I$ such that

$$[\![y \,\varepsilon\, B_i]\!] \leq \max\{ \operatorname{hgt}(A_i \cap \overline{A_k}), \min\{ \operatorname{hgt}(A_i), [\![y \,\varepsilon\, B_k]\!] \} \}.$$

Therefore, to get (4.47) true, it now suffices always to have

$$[\![y \,\varepsilon\, B_i]\!] \leq \min\{ \operatorname{hgt}(A_i), [\![y \,\varepsilon\, B_k]\!] \} \quad \text{for some } k \in I, \tag{4.51}$$

and obviously for (4.51) one now has condition (iv) as a sufficient one.

Finally, assume $t_1 = t_2 = \min$. In that case (4.47) and (4.50) become: for all $i \in I$ and all $y \in \mathcal{Y}$ there exists some $k \in I$ such that

$$\begin{aligned} [\![y \,\varepsilon\, B_i]\!] &\leq \sup_{x \in \mathcal{X}} [\![x \,\varepsilon\, A_i \wedge_t ((x \,\varepsilon\, A_k \wedge y \,\varepsilon\, B_k) \vee x \,\varepsilon\, \overline{A_k})]\!] \\ &= \sup_{x \in \mathcal{X}} [\![(x \,\varepsilon\, A_i \wedge_t (x \,\varepsilon\, A_k \wedge y \,\varepsilon\, B_k)) \vee (x \,\varepsilon\, A_i \wedge_t x \,\varepsilon\, \overline{A_k})]\!] \\ &= \max\{ \operatorname{hgt}(A_i \cap_t \overline{A_k}), \sup_{x \in \mathcal{X}} [\![x \,\varepsilon\, A_i \wedge_t (x \,\varepsilon\, A_k \wedge y \,\varepsilon\, B_k)]\!] \} \end{aligned}$$

To have this condition true it suffices if for all $i \in I$ one has

$$\mathrm{hgt}\,(B_i) \le \mathrm{hgt}\,(A_i \cap_t \overline{A_k}) \quad \text{for some } k \in I,$$

which easily gives condition (iii) for the present case. Again now assuming furthermore that also $t = \min$ is the case yields from (4.47) the condition: for all $i \in I$ and all $y \in \mathcal{Y}$ there exists some $k \in I$ such that

$$\begin{aligned} [\![y \,\varepsilon\, B_i]\!] &\le \sup_{x \in \mathcal{X}} [\![(x \,\varepsilon\, A_i \cap A_k \wedge y \,\varepsilon\, B_k) \vee (x \,\varepsilon\, A_i \cap \overline{A_k})]\!] \\ &= \max\{\min\{\mathrm{hgt}\,(A_i \cap A_k), [\![y \,\varepsilon\, B_k]\!]\}, \mathrm{hgt}\,(A_i \cap \overline{A_k})\}. \end{aligned}$$

For this inequality it suffices that for each $i \in I$:

$$\mathrm{hgt}\,(B_i) \le \mathrm{hgt}\,(A_i \cap A_k) \quad \text{for some } k \in I.$$

Thus, (iv) is also a sufficient condition in this situation. QED

As in the previous subsection 4.2.2 now the sufficient conditions for the superset property of R are not hard to meet by a fuzzy controller either, and in general conditions (i) to (iii) seem to be much more important than condition (iv) in a normal application. The situation we met in that subsection 4.2.2, that sufficient conditions for the subset property of R are much more difficult to find and to have realized, is repeated for the present situation. Furthermore, we now have to restrict ourselves to some special cases for the choice of the t-norms t_1, t_2 to find results which do not look extremely unmanageable. But, the following sufficient conditions for the subset property are also quite strong: so for example in items (i) as well as (ii) of the following proposition one has to fulfill two simultaneous demands, the first of them formulating a kind of "almost pairwise disjointness" of the input fuzzy sets, and the second one demanding some "considerable overlap" of the output fuzzy sets. That indicates that the coding procedure we are actually discussing for the single control rules is not well suited for the subset property of R.

Proposition 4.8 *Sufficient conditions for R to have the subset property are:*

(*i*) *if one assumes $t_1 = t_2 = t_L$, then*

$$1 - \mathrm{hgt}(\overline{B_i}) \ge \mathrm{hgt}(A_i \cap_t \overline{A_j}) \quad \textit{for all } i, j \in I$$

together with

$$Y^{[\mathrm{hgt}(A_i)]} \cap_{\boldsymbol{t}} B_j \subset B_i \quad \textit{for all } i,j \in I;$$

(*ii*) *if one assumes* $\boldsymbol{t} = \boldsymbol{t}_1 = \boldsymbol{t}_2 = \min$, *then*

$$1 - \mathrm{hgt}(\overline{B_i}) \geq \mathrm{hgt}(A_i \cap \overline{A_j}) \quad \textit{for all } i,j \in I$$

together with

$$Y^{[\mathrm{hgt}(A_i \cap A_j)]} \cap B_j \subset B_i \quad \textit{for all } i,j \in I.$$

Proof. Without any restrictions concerning the t-norms involved one has by (4.33) as a necessary and sufficient condition for the subset property of R that for all $i,j \in I$ and all $y \in \mathcal{Y}$

$$\begin{aligned} [\![y \varepsilon B_i]\!] &\geq [\![\exists x (x \varepsilon A_i \wedge_{\boldsymbol{t}} ((x,y) \varepsilon \Upsilon(A_j \models\!\!\!\!\Rightarrow B_j)))]\!] \\ &= \sup_{x \in \mathcal{X}} [\![x \varepsilon A_i \wedge_{\boldsymbol{t}} ((x,y) \varepsilon (A_j \times_{\boldsymbol{t}_1} B_j) \cup_{\boldsymbol{t}_2} (\overline{A_j} \times_{\boldsymbol{t}_1} Y)]\!] \\ &= \sup_{x \in \mathcal{X}} [\![x \varepsilon A_i \wedge_{\boldsymbol{t}} ((x \varepsilon A_j \wedge_{\boldsymbol{t}_1} y \varepsilon B_j) \vee_{\boldsymbol{t}_2} x \varepsilon \overline{A_j})]\!]. \end{aligned} \tag{4.52}$$

For $\boldsymbol{t}_1 = \boldsymbol{t}_2 = \boldsymbol{t}_L$ this becomes as in the last proof

$$\begin{aligned} [\![y \varepsilon B_i]\!] &\geq \sup_{x \in \mathcal{X}} [\![x \varepsilon A_i \wedge_{\boldsymbol{t}} (x \varepsilon \overline{A_j} \vee y \varepsilon B_j)]\!] \\ &= \sup_{x \in \mathcal{X}} [\![(x \varepsilon A_i \cap_{\boldsymbol{t}} \overline{A_j}) \vee (x \varepsilon A_i \wedge_{\boldsymbol{t}} y \varepsilon B_j)]\!] \\ &= \max\{\mathrm{hgt}\,(A_i \cap_{\boldsymbol{t}} \overline{A_j}), \mathrm{hgt}\,(A_i) \wedge_{\boldsymbol{t}} [\![y \varepsilon B_j]\!]\}. \end{aligned} \tag{4.53}$$

Thus a sufficient condition results if both values, the maximum of which is taken in (4.53), separately are no greater than the left-hand side $[\![y \varepsilon B_i]\!]$. For $\mathrm{hgt}\,(A_i \cap_{\boldsymbol{t}} \overline{A_j})$ this worst case scenario demands that one has $1 - \mathrm{hgt}\,(\overline{B_i}) \geq \mathrm{hgt}\,(A_i \cap_{\boldsymbol{t}} \overline{A_j})$ for all $i,j \in I$. And in the other case the demand reads $[\![y \varepsilon B_i]\!] \geq \mathrm{hgt}\,(A_i) \wedge_{\boldsymbol{t}} [\![y \varepsilon B_j]\!]$, which holds true for $Y^{[\mathrm{hgt}\,(A_i)]} \cap_{\boldsymbol{t}} B_j \subset B_i$. Thus we have proven case (i).

Starting from $\boldsymbol{t} = \boldsymbol{t}_1 = \boldsymbol{t}_2 = \min$ one gets the condition that for all $i,j \in I$ and all $y \in \mathcal{Y}$

$$\begin{aligned} [\![y \varepsilon B_i]\!] &\geq \sup_{x \in \mathcal{X}} [\![(x \varepsilon A_i \cap \overline{A_j}) \vee (x \varepsilon A_i \cap A_j \wedge y \varepsilon B_j)]\!] \\ &= \max\{\mathrm{hgt}\,(A_i \cap \overline{A_j}), \min\{\mathrm{hgt}\,(A_i \cap A_j), [\![y \varepsilon B_j]\!]\}\}. \end{aligned}$$

The same arguments used to infer conditions (i) from (4.53) now give conditions (ii). QED

4.2.4 Coding rules by $\Upsilon(A \mapsto B) = (\overline{A} \times_{\mathbf{t}_1} Y) \cup_{\mathbf{t}_2} (X \times_{\mathbf{t}_1} B)$

The coding procedure for the single control rules now can be understood as reading the fuzzy implication (4.45) as

either INPUT $= \overline{A}$ **or** OUTPUT $= B$

which is closely analogous to the standard way of classical propositional logic to express implication by disjunction and negation.

Proposition 4.9 *Sufficient conditions for R to have the superset property are:*

(*i*) *without restrictions concerning the t-norms involved:*

$$\mathrm{hgt}(A_i) = 1 \quad \textit{for all } i \in I;$$

(*ii*) *in the case that the t-norm $\mathbf{t}$ is distributive over the t-conorm $\mathbf{s}_{\mathbf{t}_2}$, thus for example in the case that $\mathbf{t}_2 = \min$,*

$$\mathrm{hgt}(B_i) \leq \mathrm{hgt}(A_i \cap_{\mathbf{t}} \bigcup_{j \in I} \overline{A_j}) \quad \textit{for all } i \in I;$$

(*iii*) *in the case that $\mathbf{t} = \mathbf{t}_2 = \mathbf{t}_L$: for each $i \in I$ there exists some $k \in I$ such that*

$$A_i^{\geq \mathrm{hgt}(B_i)} \cap (A_i \cap_{\mathbf{t}_L} \overline{A_j})^{\geq \mathrm{hgt}(B_i)} \neq \emptyset.$$

Proof. By (4.32) we also have as a necessary and sufficient condition for the superset property in the present case that for all $i \in I$ and all $y \in \mathcal{Y}$ there exists some $k \in I$ such that

$$\begin{aligned} [\![y \,\varepsilon\, B_i]\!] &\leq [\![\exists x (x \,\varepsilon\, A_i \wedge_{\mathbf{t}} ((x,y) \,\varepsilon\, \Upsilon(A_k \mapsto B_k)))]\!] \\ &= \sup_{x \in \mathcal{X}} [\![x \,\varepsilon\, A_i \wedge_{\mathbf{t}} ((x,y) \,\varepsilon\, (\overline{A_k} \times_{\mathbf{t}_1} Y) \cup_{\mathbf{t}_2} (X \times_{\mathbf{t}_1} B_k)]\!] \\ &= \sup_{x \in \mathcal{X}} [\![x \,\varepsilon\, A_i \wedge_{\mathbf{t}} (x \,\varepsilon\, \overline{A_k} \vee_{\mathbf{t}_2} y \,\varepsilon\, B_k)]\!]. \end{aligned} \tag{4.54}$$

As a side remark let us mention that the t-norm $\mathbf{t}_1$ does not appear any more in the final formula (4.54), hence the choice of $\mathbf{t}_1$ is inessential for the present coding procedure of the single control rules.

Using in (4.54) the monotonicity of $\mathbf{s}_{\mathbf{t}_2}$ together with $[\![x \,\varepsilon\, \overline{A_k}]\!] \geq 0$ gives as a sufficient condition that for all $i \in I$ one has

$$\begin{aligned} [\![y \,\varepsilon\, B_i]\!] &\leq \sup_{x \in \mathcal{X}} [\![x \,\varepsilon\, A_i \wedge_{\mathbf{t}} y \,\varepsilon\, B_k]\!] \\ &= \mathrm{hgt}\,(A_i)\, \mathbf{t}\, [\![y \,\varepsilon\, B_k]\!] \end{aligned}$$

for some $k \in I$. Hence (i) is a sufficient condition.

With the distributivity assumption of (ii) one gets from (4.54) the condition that for all $i \in I$ and all $y \in \mathcal{Y}$ there exists some $k \in I$ such that

$$\begin{aligned}[][\![y \,\varepsilon\, B_i]\!] &\leq \sup_{x\in\mathcal{X}}[\![(x \,\varepsilon\, A_i \wedge_{\mathbf{t}} x \,\varepsilon\, \overline{A_k}) \vee_{\mathbf{t}_2} (x \,\varepsilon\, A_i \wedge_{\mathbf{t}} y \,\varepsilon\, B_k)]\!] \\ &= \sup_{x\in\mathcal{X}}[\![(x \,\varepsilon\, A_i \cap_{\mathbf{t}} \overline{A_k}) \vee_{\mathbf{t}_2} \ldots)]\!]\end{aligned}$$

and thus immediately condition (ii) because of

$$\operatorname{hgt}(A_i \cap_{\mathbf{t}} \overline{A_k}) \leq \sup_{x\in\mathcal{X}}[\![(x \,\varepsilon\, A_i \cap_{\mathbf{t}} \overline{A_k}) \vee_{\mathbf{t}_2} \ldots)]\!]$$

for all $i, k \in I$.

And in the case that $\mathbf{t} = \mathbf{t}_2 = \mathbf{t}_L$, using directly the definitions of $\mathbf{t}_L, \mathbf{s}_{\mathbf{t}_L}$ given in Tables 1.1 and 1.2 one has from (4.54) the condition that for all $i \in I$ and all $y \in \mathcal{Y}$ there exists some $k \in I$ such that

$$\begin{aligned}[][\![y \,\varepsilon\, B_i]\!] &\leq \sup_{x\in\mathcal{X}} \max\{0, [\![x \,\varepsilon\, A_i]\!] + \min\{[\![x \,\varepsilon\, \overline{A_k}]\!] + [\![y \,\varepsilon\, B_k]\!], 1\} - 1\} \\ &= \sup_{x\in\mathcal{X}} \max\{0, \min\{[\![x \,\varepsilon\, A_i]\!] + [\![x \,\varepsilon\, \overline{A_k}]\!] - 1 + [\![y \,\varepsilon\, B_k]\!], [\![x \,\varepsilon\, A_i]\!]\}\} \\ &= \max\{0, \sup_{x\in\mathcal{X}} \min\{[\![x \,\varepsilon\, A_i \,\&\, x \,\varepsilon\, \overline{A_k}]\!] + [\![y \,\varepsilon\, B_k]\!], [\![x \,\varepsilon\, A_i]\!]\}\}. \qquad (4.55)\end{aligned}$$

To have this condition true it is surely sufficient that the final term of (4.55) is no smaller than $\operatorname{hgt}(B_i)$. And condition (iii) of our proposition yields just this. QED

Proposition 4.10 *Sufficient conditions for R to have the subset property are:*

(i) if one assumes $\mathbf{t}_2 = \min$ then

$$1 - \operatorname{hgt}(\overline{B_i}) \geq \operatorname{hgt}(A_i \cap_{\mathbf{t}} \overline{A_j}) \quad \textit{for all } i, j \in I$$

together with

$$Y^{[\operatorname{hgt}(A_i)]} \cap_{\mathbf{t}} B_j \subset B_i \quad \textit{for all } i, j \in I;$$

(ii) if one assumes $\mathbf{t} = \mathbf{t}_2 = \mathbf{t}_L$ then

$$1 - \operatorname{hgt}(\overline{B_i}) \geq \operatorname{hgt}(A_i) \quad \textit{for all } i, j \in I.$$

Proof. Without any restrictions concerning the t-norms involved one now has by (4.33) as a necessary and sufficient condition for the subset property of R that for all $i, j \in I$ and all $y \in \mathcal{Y}$

$$[\![y \,\varepsilon\, B_i]\!] \geq \sup_{x\in\mathcal{X}} [\![x \,\varepsilon\, A_i \wedge_{\mathbf{t}} (x \,\varepsilon\, \overline{A_j} \vee_{\mathbf{t}_2} y \,\varepsilon\, B_j)]\!]. \tag{4.56}$$

Starting from (4.56) and using $\mathbf{t}_2 = \min$ one gets the condition that for all $i, j \in I$ and all $y \in \mathcal{Y}$

$$\begin{aligned} [\![y \,\varepsilon\, B_i]\!] &\geq \sup_{x\in\mathcal{X}} [\![(x \,\varepsilon\, A_i \cap_{\mathbf{t}} \overline{A_j}) \vee (x \,\varepsilon\, A_i \wedge_{\mathbf{t}} y \,\varepsilon\, B_j)]\!] \\ &= \max\{\mathrm{hgt}\,(A_i \cap_{\mathbf{t}} \overline{A_j}), \mathrm{hgt}\,(A_i)\, \mathbf{t}\, [\![y \,\varepsilon\, B_j]\!]\}. \end{aligned}$$

To get a sufficient condition out of these inequalities we again take the worst case scenario for the left hand side and try to have the last-mentioned right-hand side $\leq 1 - \mathrm{hgt}\,(\overline{B_i})$, i.e. to have both terms, the maximum is taken over there, to be $\leq 1 - \mathrm{hgt}\,(\overline{B_i})$. This immediately gives both parts of (i).

Finally, in the case that $\mathbf{t} = \mathbf{t}_2 = \mathbf{t}_L$ the same calculations as in the last proof give, from (4.56), the necessary and sufficient condition that for all $i, j \in I$ and all $y \in \mathcal{Y}$

$$[\![y \,\varepsilon\, B_i]\!] \geq \max\{0, \sup_{x\in\mathcal{X}} \min\{[\![x \,\varepsilon\, A_i \,\&\, x \,\varepsilon\, \overline{A_k}]\!] + [\![y \,\varepsilon\, B_k]\!], [\![x \,\varepsilon\, A_i]\!]\}\}.$$

This obviously is equivalent to the simpler condition that for all $i, j \in I$ and all $y \in \mathcal{Y}$ one has

$$[\![y \,\varepsilon\, B_i]\!] \geq \min\{[\![x \,\varepsilon\, A_i \,\&\, x \,\varepsilon\, \overline{A_j}]\!] + [\![y \,\varepsilon\, B_j]\!], [\![x \,\varepsilon\, A_i]\!]\}.$$

The same worst case scenario as just before, now applied with respect to the second term within the min-operator, immediately gives (ii). QED

4.3 Solvability degrees and the manipulation of fuzzy data

Let us start with the idea that the fuzzy data we are considering for the input and output values in our fuzzy model themselves to some extent are given only approximately. This, of course, may be caused by quite different reasons. One such possibility, sometimes realized in engineering practice, is that one has a relatively clear idea of higher membership degrees, but much less information concerning small membership degrees. Another possibility is that some noise – with small values – makes the smaller membership degrees (especially) uncertain. A third possibility may be realized in case one first has a linguistic model of some process and then has to "translate" the linguistic values into fuzzy sets: in this case one also has some degree of freedom for the actual choice of these fuzzy inputs and outputs.

One of the types of approaches often considered in the case of such a kind of uncertainty concerning the fuzzy "data" is to change from usual fuzzy sets to fuzzy sets of type 2 or perhaps only to interval valued fuzzy sets. We shall not follow this strategy.

Instead here we assume that some *threshold level* α is given such that membership degrees smaller than α are less reliable than those greater than this threshold α.

To discuss the influence of such a threshold level on the existence of solutions of fuzzy relation equations and of systems of such equations let us first introduce some additional notation.

Let $\alpha \in [0,1]$ be a threshold level. For a given fuzzy set $A \in I\!F(\mathcal{X})$ define

$$A^{\alpha} =_{def} A \cup X^{[\alpha]}, \tag{4.57}$$

i.e. put for the membership degrees

$$\mu_{A^{\alpha}}(x) = \max\{\mu_A(x), \alpha\} = \begin{cases} \mu_A(x) & \text{if } \mu_A(x) \geq \alpha \\ \alpha & \text{if } \mu_A(x) < \alpha, \end{cases}$$

and correspondingly define

$$A_{\alpha} =_{def} A \cap \bigcup_{a \in A^{\geq \alpha}} \langle\!\langle a \rangle\!\rangle_1, \tag{4.58}$$

which means for the membership degrees

$$\mu_{A_{\alpha}}(x) = \begin{cases} \mu_A(x) & \text{if } \mu_A(x) \geq \alpha \\ 0 & \text{if } \mu_A(x) < \alpha. \end{cases}$$

Both fuzzy sets A^{α}, A_{α} are borderline cases of those fuzzy sets which correspond to the given fuzzy set A according to the threshold level α and its interpretation.

In the following, we use the threshold level to change our fuzzy relation equations. Instead of equations of the type

$$A \circ_t R = B \tag{4.59}$$

we consider equations of the kinds

$$A^{\alpha} \circ_t R = B^{\alpha} \qquad \text{and} \qquad A_{\alpha} \circ_t R = B_{\alpha}. \tag{4.60}$$

In the same way we change the systems (4.28) of fuzzy relation equations. To discuss the change in solvability behaviour connected with these changes of fuzzy relation equations we consider the corresponding solvability degrees.

Before coming back to systems of fuzzy relation equations let us first consider the case that we will – according to our discussions related to the threshold level – change the input and output data A, B of fuzzy relation equations (4.59) to A^α, B^α or to A_α, B_α. Let us denote the corresponding solvability degrees by $\xi(\alpha)$ and by $\xi'(\alpha)$, i.e. according to proposition 3.21 we consider

$$\xi(\alpha) =_{def} \operatorname{hgt}(B^\alpha) \varphi_{\mathbf{t}} \operatorname{hgt}(A^\alpha), \tag{4.61}$$

$$\xi'(\alpha) =_{def} \operatorname{hgt}(B_\alpha) \varphi_{\mathbf{t}} \operatorname{hgt}(A_\alpha). \tag{4.62}$$

Our problem will be to consider the relations of the threshold depending *relative* solvability degrees $\xi(\alpha), \xi'(\alpha)$ to the "absolute" solvability degree ξ_0, i.e. to consider the change in the solvability behaviour of (4.59) caused by the changes of input and output data motivated by the threshold level α.

Proposition 4.11 *Let* $\mathbf{t}$ *be a continuous t-norm and* α, β *threshold levels. Then for the corresponding solvability degrees of equation (4.59) there hold true*

(*i*) $\alpha \leq \xi(\alpha) \leq 1$,

(*ii*) $\xi_0 \leq \xi(\alpha)$,

(*iii*) $\xi(0) = \xi_0$ *and* $\xi(1) = 1$,

(*iv*) $\alpha \leq \beta \Rightarrow \xi(\alpha) \leq \xi(\beta)$.

Proof. (i) Obviously $\xi(\alpha) \leq 1$. And the worst case, i.e. smallest value of $\xi(\alpha)$ is given, according to (4.61), in the case that $\operatorname{hgt}(B^\alpha) = 1$ and, because of $\operatorname{hgt}(A^\alpha) \geq \alpha$, that $\operatorname{hgt}(A^\alpha) = \alpha$. But then $\xi(\alpha) = 1 \varphi_{\mathbf{t}} \alpha = \alpha$ results.

(ii) Immediately one has $\operatorname{hgt}(A^\alpha) = \max\{\operatorname{hgt}(A), \alpha\}$ and analogously for B^α. Hence

$$\operatorname{hgt}(B) \leq \operatorname{hgt}(A) \;\Rightarrow\; \operatorname{hgt}(B^\alpha) \leq \operatorname{hgt}(A^\alpha),$$

$$\operatorname{hgt}(B), \operatorname{hgt}(A) \leq \alpha \;\Rightarrow\; \operatorname{hgt}(B^\alpha) = \operatorname{hgt}(A^\alpha),$$

i.e. $\xi(\alpha) = 1$ in both cases, and finally

$$\operatorname{hgt}(B) > \max\{\operatorname{hgt}(A), \alpha\} \Rightarrow$$

$$\operatorname{hgt}(B^\alpha) = \operatorname{hgt}(B) \quad \text{and} \quad \operatorname{hgt}(A^\alpha) \geq \operatorname{hgt}(A)$$

such that $\xi_0 \leq \xi(\alpha)$ in this case, by monotonicity of φ_t in the second argument. Hence $\xi_0 \leq \xi(\alpha)$ in general.

(iii) Of course, $A^0 = A$ and $B^0 = B$, thus $\xi(0) = \xi_0$. And if we consider the 1-universal fuzzy subset $X^{[1]}$ of the universe of discourse $\mathcal{X}$ whose membership degree is =1 in each point of that universe, then $A^1 = X^{[1]}$, as well as $B^1 = Y^{[1]}$, and hence $\xi(1) = 1$ because of $X^{[1]} \circ_t (X^{[1]} \times_t Y^{[1]}) = Y^{[1]}$ and the fact that $X^{[1]} \times_t Y^{[1]}$ is the 1-universal fuzzy subset of the universe of discourse $\mathcal{X} \times \mathcal{Y}$.

(iv) Interpret A^α, B^α as the original input and output data for the fuzzy relation equation (4.59) which will be modified according to a threshold level β. Then from $\alpha \leq \beta$ one gets $(A^\alpha)^\beta = A^\beta$ and $(B^\alpha)^\beta = B^\beta$. Now apply (ii).

QED

Thus we know that the solvability behaviour of equation (4.59) becomes better and better, i.e. the solvability degree becomes greater and greater with growing threshold level α.

From an intuitive point of view it seems that the idea connected with the introduction of threshold level α may not only be realized by the data transformation

$$A \mapsto A^\alpha \qquad \text{and} \qquad B \mapsto B^\alpha \tag{4.63}$$

but equally well by the corresponding data transformation

$$A \mapsto A_\alpha \qquad \text{and} \qquad B \mapsto B_\alpha \tag{4.64}$$

However, this is not the case. For example the solvability degree $\xi'(\alpha)$ from (4.62) has not the monotonicity property analogous to proposition 4.11 (iv). To show this, we consider a concrete example for two different kinds of t-norms.

Let us consider the following two cases and assume that A, B are fuzzy singletons with

$$\operatorname{hgt}(A) = 0.7 \qquad \text{and} \qquad \operatorname{hgt}(B) = 0.9.$$

Case 1. Take as t-norm $t = t_G = \min$ with the corresponding Φ-operator φ_G satisfying according to Table 1.3

$$u \, \varphi_G \, v = \begin{cases} 1, & \text{if } u \leq v \\ v, & \text{if } u > v. \end{cases} \tag{4.65}$$

Case 2. Take as t-norm $t = t_L = \&$ as given in Table 1.1 with the corresponding Φ-operator φ_L satisfying according to Table 1.3

$$u \, \varphi_L \, v = \min\{1, 1 - u + v\},$$

which is the well-known Łukasiewicz implication operator of many-valued logic.

The values of $\xi'(\alpha)$ for different threshold levels are collected in the following Table 4.1.

Threshold	Case 1	Case 2
$\alpha \leq 0.7$	$\xi'(\alpha) = 0.7$	$\xi'(\alpha) = 0.8$
$0.7 < \alpha \leq 0.9$	$\xi'(\alpha) = 0$	$\xi'(\alpha) = 0.1$
$\alpha > 0.9$	$\xi'(\alpha) = 1$	$\xi'(\alpha) = 1$

Table 4.1: Relative solvability degrees with respect to the data transformation (4.64)

Of course, smaller t-norms than $t = \&$ will give greater values of $\xi'(\alpha)$, but the qualitative behaviour will be still the same: a much smaller value of $\xi'(\alpha)$ for $0.7 < \alpha \leq 0.9$ as for $\alpha \leq 0.7$ and $\alpha > 0.9$. And it is this effect that makes (for the author's feeling) the relative solvability degree $\xi'(\alpha)$ inappropriate for the present purposes.

Therefore, for systems (4.28) of fuzzy relation equations we will discuss only the data transformations of kind (4.63) and the corresponding changes $\xi \mapsto \xi(\alpha)$ of solvability degrees.

Finally, let us discuss the influence that has a data transformation à la (4.63) for the solvability degree of system (4.28). We denote by $\xi(\alpha)$ now the solvability degree of the system

$$A_i^\alpha \circ_t R = B_i^\alpha, \qquad i = 1, \ldots, n,$$

i.e. the (relative) solvability degree of the "transformed system" of (4.28) with respect to the threshold level α. Unfortunately, again, the lack of a direct characterization of the systems solvability degree ξ allows us to prove only some of the properties that correspond to the results of the previous proposition 4.11.

Proposition 4.12 *Let* $\mathbf{t}$ *be a continuous t-norm,* α *any threshold level. Then we have for the solvability degrees of the modified systems* (4.28) *of fuzzy equations:*

$$(i) \quad \xi(0) = 0 \quad \text{and} \quad \xi(1) = 1,$$

$$(ii) \quad \mathop{\mathbf{T}}_{i=1}^{n} (\alpha \,\mathbf{t}\, \alpha) \leq \xi(\alpha) \leq 1,$$

$$(iii) \quad \text{if all input fuzzy sets are normal then } \mathop{\mathbf{T}}_{i=1}^{n} \alpha \leq \xi(\alpha).$$

Proof. (i) and $\xi(\alpha) \leq 1$ in (ii) are obvious. From proposition 3.24 we get

$$\xi(\alpha) \geq \mathop{\mathbf{T}}_{i=1}^{n} \bigwedge_{y \in \mathcal{Y}} (1 \,\varphi_{\mathbf{t}}\, (\alpha \,\mathbf{t}\, \alpha)) = \mathop{\mathbf{T}}_{i=1}^{n} (\alpha \,\mathbf{t}\, \alpha)$$

which completes (ii). From proposition 3.24 we also get

$$\xi(\alpha) \geq \mathop{\mathbf{T}}_{i=1}^{n} \bigwedge_{y \in \mathcal{Y}} (1 \,\varphi_{\mathbf{t}}\, (1 \,\mathbf{t}\, \alpha)) = \mathop{\mathbf{T}}_{i=1}^{n} \alpha$$

supposing the extra assumption of (iii). Thus also (iii). QED

Again, the case $\mathbf{t} = \wedge = \min$ with its direct characterization of the solvability degree allows us to prove more.

Proposition 4.13 *Let* $\mathbf{t} = \wedge$ *and* α, β *be any threshold levels. Then there hold true*

$$(i) \quad \alpha \leq \xi(\alpha) \leq 1,$$

$$(ii) \quad \xi \leq \xi(\alpha),$$

$$(iii) \quad \alpha \leq \beta \Rightarrow \xi(\alpha) \leq \xi(\beta).$$

Proof. (i) comes directly from proposition 4.12 (ii).

For (ii) remember formula (3.24). As a first remark note that always

$$\mu_{A_j}(x) \,\varphi_G\, \mu_{B_j}(y) \leq \mu_{A_j^\alpha}(x) \,\varphi_G\, \mu_{B_j^\alpha}(y) \tag{4.66}$$

because for $\mu_{A_j}(x) \leq \alpha$ we have the value 1 on the right hand side, and in case $\mu_{A_j}(x) > \alpha$ we have (4.66) because of $\mu_{B_j}(y) \leq \mu_{B_j^\alpha}(y)$. Furthermore because of $\alpha \leq \mu_{B_j^\alpha}(y)$ and (4.65): $\alpha \leq \mu_{A_j^\alpha}(x) \varphi_G \mu_{B_j^\alpha}(y)$. Hence

$$\bigvee_{x \in \mathcal{X}} \min\left\{\mu_{A_i}(x), \bigwedge_{1 \leq j \leq n} (\mu_{A_j}(x) \varphi_G \mu_{B_j}(y))\right\}$$
$$\leq \bigvee_{x \in \mathcal{X}} \min\left\{\mu_{A_i^\alpha}(x), \bigwedge_{1 \leq j \leq n} (\mu_{A_j^\alpha}(x) \varphi_G \mu_{B_j^\alpha}(y))\right\}$$

and also

$$\alpha \leq \bigvee_{x \in \mathcal{X}} \min\left\{\mu_{A_i^\alpha}(x), \bigwedge_{1 \leq j \leq n} (\mu_{A_j^\alpha}(x) \varphi_G \mu_{B_j^\alpha}(y))\right\}.$$

Thus by the same arguments we used to establish (4.66) now we get for each index i

$$\mu_{B_i}(y) \varphi_G \bigvee_{x \in \mathcal{X}} \min\left\{\mu_{A_i}(x), \bigwedge_{1 \leq j \leq n} (\mu_{A_j}(x) \varphi_G \mu_{B_j}(y))\right\}$$
$$\leq \mu_{B_i}(y) \varphi_G \bigvee_{x \in \mathcal{X}} \min\left\{\mu_{A_i^\alpha}(x), \bigwedge_{1 \leq j \leq n} (\mu_{A_j^\alpha}(x) \varphi_G \mu_{B_j^\alpha}(y))\right\}$$

and from this (ii).

Finally, (iii) is a consequence of (ii) in the same way as in proposition 4.11 item (iv) was a consequence of item (ii) there: simply use the same argument. QED

For "small" values of the threshold level α one can assume that the data transformation (4.63) – as well as the data transformation (4.64) – preserves the most significant parts of the shapes of the membership functions, because only membership degrees lower than α are changed. Therefore, in a calculation of a nonfuzzy value for the control output $y_0 \in \mathcal{Y}$, for example according to the method of the centre of gravity as introduced in MAMDANI (1976), one should neglect all the values of the membership functions B_i^α smaller than α, i.e. one should take in the case of an output value $B := R''A = A \circ R$ with respect to a concrete input value A:

$$y_0 = \sum_{\{y \in \mathcal{Y} \mid \mu_B(y) > \alpha\}} y \cdot \mu_B(y) \Bigg/ \sum_{\{y \in \mathcal{Y} \mid \mu_B(y) > \alpha\}} \mu_B(y) \tag{4.67}$$

in the case of finite universes of discourse $\mathcal{X}, \mathcal{Y}$, and one should – together with the assumptions of the existence of the integrals to be considered – take

$$y_0 = \int_{\{y \in \mathcal{Y} \mid \mu_B(y) > \alpha\}} y \cdot \mu_B(y)\, dy \Bigg/ \int_{\{y \in \mathcal{Y} \mid \mu_B(y) > \alpha\}} \mu_B(y)\, dy. \tag{4.68}$$

for continuous universes of discourse $\mathcal{X}, \mathcal{Y}$.

The method for the handling of fuzzy data we just have discussed, shall now be illustrated by a numerical example of an industrial-oriented system from MAMDANI/ASSILIAN (1975). Here we sketch only the background of the technical system, referring for details to the original paper.

The linguistic rules describe a strategy of control for a steam engine. The number of control rules is 15, while the knowledge base contains statements of the following type:

$$\text{if } \text{INPUT} = A_i \ (= E_i \times CE_i) \quad \text{then } \text{OUTPUT} = B_i, \qquad i = 1, \dots, 15. \tag{4.69}$$

Here in the i-th rule the input value $A_i = E_i \times CE_i$ is the fuzzy cartesian product of the fuzzy description of the *error* (E_i) and the *change of error* (CE_i) with respect to the intended state of the steam engine, that means

$$\mu_{A_i}(x, y) = \mu_{E_i}(x) \wedge \mu_{CE_i}(y),$$

and B_i is the fuzzy set of control of that i-th rule.

From the rules (4.69) the fuzzy relation R is computed according to the MAMDANI strategy, i.e. according to formula (4.23), and therefore one has

$$R = \bigcup_{i=1}^{15} (A_i \times B_i) = \bigcup_{i=1}^{15} ((E_i \times CE_i) \times B_i).$$

We consider only the t-norm $t = \min$ and get with the inference mechanism given by $(E_i \times CE_i) \circ R$ for the system of fuzzy relation equations corresponding to (4.69) the solvability degree

$$\xi(\alpha) = \bigwedge_{i=1}^{15} \xi_i(\alpha)$$

where $\xi_i(\alpha)$ denotes the solvability degree of the fuzzy relation equation for the i-th rule. The results are shown in Figure 4.1. Thereby α takes discrete values varying from 0.0 to 1.0 with a discretization step of 0.1.

Note that the value of the threshold level equal to 0.3 still yields the value 1 for the solvability index $\xi(0.3)$. Moving down with threshold α we obtain lower values of this solvability index for the entire set of control rules. But the information which the (relative) solvability degree of the whole system provides is quite global. What is lacking with this degree $\xi(\alpha)$ is information about the detailed behaviour of the single rules. Thus one also has to look

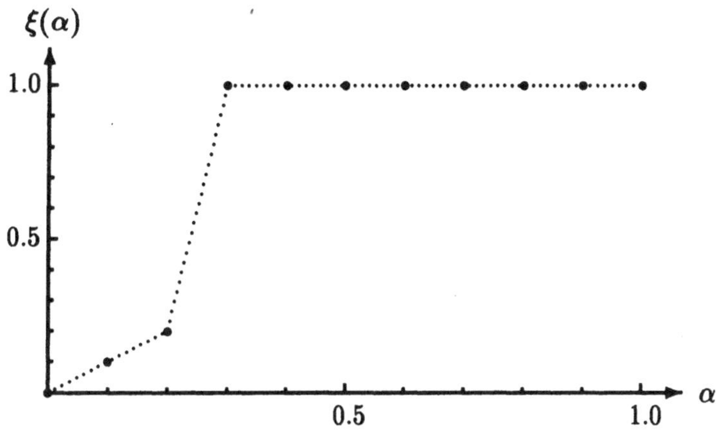

Figure 4.1: Solvability degree $\xi(\alpha)$ of the MAMDANI/ASSILIAN-system (4.69) as a function of the threshold level α.

for the (relative) solvability degrees of the single rules. Below in Table 4.2 we summarize for the MAMDANI/ASSILIAN-system the rules which have the lowest values of the solvability index $\xi_i(\alpha)$.

Of course, decreasing values of α produce an ever-extending list of control rules that shows lower values of their degrees of solvability, in the present example especially the rules no. 6, 12, 15 – numbered according to the details in MAMDANI/ASSILIAN (1975) – are critical with respect to their (relative) solvability degrees.

Low values of the solvability degrees of some control rules are confirmed

α	number of control rule														
	1	2	3	4	5	6	7	8	9	10	11	12	13	14	15
0.2						.2						.2			.2
0.1						.1				.9		.1	.1	.1	.1
0			.5			0			.5	.9		0	0	0	0

Table 4.2: Relative solvability degrees $\xi_i(\alpha)$ of the MAMDANI/ASSILIAN-controller (4.69)[10]

by another construction that evaluates their credibility in sense of a certainty factor CF, cf. PEDRYCZ (1985). Then in this example for the 12th control rule this certainty factor has the value $CF_{12} = 0$. The rules with numbers 6, 13, 14, 15 have values of the certainty factor CF which are below 0.5. Indeed, in the above construction for $\alpha = 0.2$, one correspondingly has a subset of three control rules: no. 6, 12, 15, for which the values $\xi_i(0.2)$ do not exceed 0.2.

For the threshold value $\alpha_0 = 0.3$ some of the output fuzzy sets B_i^α, i.e. of the fuzzy control advices, have been displayed in Figure 4.2.

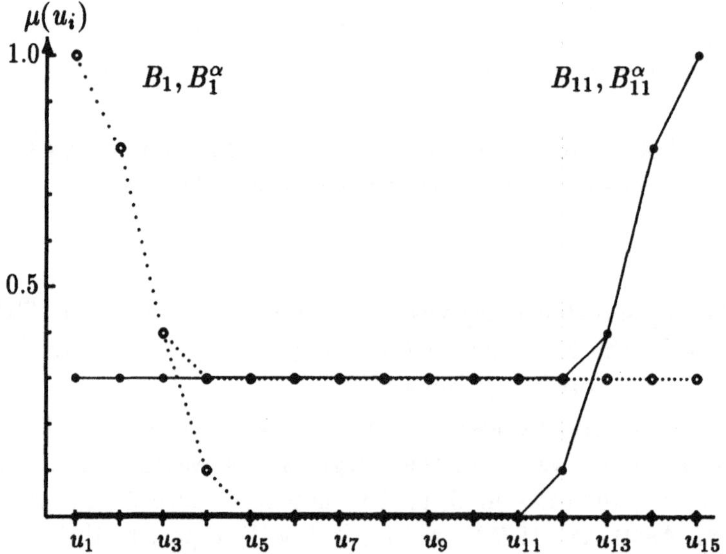

Figure 4.2: Fuzzy output sets B_i and B_i^α for $i = 1, 11$ and $\alpha = 0.3$

The value $\alpha_0 = 0.3$ is low enough such that the changes $A_i \mapsto A_i^{0.3}$ and $B_i \mapsto B_i^{0.3}$, which according to our method will yield a completely solvable system of relation equations, i.e. a non-interactive fuzzy controller out of the system (4.69) of control rules, will not hide the structure of the relation of the model. This structure can essentially be seen in the output fuzzy sets B_i – and Figure 4.2 explains that the modified outputs $B_i^{0.3}$ also present the essentials of that structure.

The approach towards the manipulation of fuzzy data that we have considered in this section enables us to use the formal apparatus from fuzzy

[10] Each empty entry means the relative solvability degree =1.

relational equations in an interesting way for the evaluation of fuzzy models. It is remarkable to underline the fact that two mechanisms, the mechanism for combining the fuzzy pieces of evidence and that one which plays with inference with the use of fuzzy information, should be discussed simultaneously, and they are significantly associated. The concrete way of combining fuzzy data leads to a certain way for inferring fuzzy consequents. These facts result from a direct association of the composition operator $\circ_t$ and the corresponding Φ-operator. The first is used in the implementation of the inference schema, while the latter is joined with the particularization of a proper way of combining the fuzzy premises. This way of thinking gives an alternative in comparison to commonly proposed schemata which come from the compositional rule of inference as stated by ZADEH.

Moreover the introduction of the solvability degree of the system of fuzzy relational equations makes it possible to express the property of "easiness" of solving that system. It provides an opportunity to look at appropriate solutions with regard to the level of threshold. We have discussed a way of changing the data set that makes it possible to render the set of fuzzy equations solvable, and moreover to measure the deformation of fuzzy data used in this process.

One additional remark of a general nature. A global identification procedure, despite the nature of the factor of uncertainty taken into account, consists of three steps:

(a) the selection of the structure of the model proposed (i. e. inputs, outputs, order of the model, etc.);

(b) the estimation of model parameters;

(c) the test of the model with respect to the data set provided (usually it is the same one as applied in the second step).

For comparison, let us look at the identification task in the presence of randomness. Then the model is constructed with the help of statistical methods. The first step relies on a choice of a "reasonable" type of function expressing a dependence between input and output variables. Further, unknown parameters of the model (the function) are estimated. Note, moreover, that at this stage the values of the parameters are computed and a level of their uncertainty is also expressed via relevant confidence intervals. Hence, even if in the real world situation the parameters of a model are nonrandom, the estimation procedure involves that they become viewed as random variables.

Finally, a consistent degree of the model is commonly performed by means of F-statistics. And if this stage provides a negative result, the entire procedure is repeated, changing the structure of the model.

The identification procedure which (at least implicitely) is proposed with the ideas of this section fits this general schema. The structure of a fuzzy model – of the type of a fuzzy controller – is established selecting the form of the fuzzy relation equations, viz. the type of the composition operation together with the t-norm involved in it, i.e. is established with the control rules and their translations. The parameter estimation for the model then proceeds through our methods for discussing the solvability of (systems of) fuzzy relation equations. Here, the solvability degrees serve as indices indicating how credible the structure of the fuzzy relation is which constitutes the fuzzy controller and hence how good the fuzzy model is. Additionally, in the methodological perspective this (relative) solvability degree has an analogous role like a confidence interval: if $\xi(\alpha)$ is lower than a borderline value $\xi^*(\alpha)$ for a certain α – then repeat the identification procedure, i.e. "adjust" the fuzzy model.

The range of the applicability of this specialized discussion via relative solvability degrees is closely related to those systems where a tool for manipulating fuzzy data is needed, e.g. expert systems, robotics, and pattern recognition.

Chapter 5

Some methodological issues of fuzzy modelling

5.1 The comparison of fuzzy sets and the inverse problem

Describing fuzzy models by their input-output behaviour is a very convenient tool in fuzzy modelling. This approach immediately leads to fuzzy relational equations – usually to systems of them – and to the necessity to solve them. Solving such systems of equation is, as the topics discussed in the previous chapters 3 and 4 indicate, a complicated task, not only for numerical reasons but also because of the fact that in the most essential cases in general only approximate solutions exist. Therefore, a detailed discussion of the solvability behaviour of systems of fuzzy relational equations and of factors which influence or change this behaviour is highly interesting, not only from a theoretical perspective but for very practical reasons of fuzzy model building too.

The choice of the fuzzy operators used in the equations, i.e. the choice of the t-norms in the definitions of $R''A$ and $R \circ_t S$, is one point for discussion. Another is the change from sup-min composition or inf-max composition, which are used within the method of getting the fuzzy output values out of the fuzzy input values, i.e. are used in the compositional rule of inference, to other, perhaps more complicated rules like those ones in (3.37) and (3.38) or even some convex combination of such simple kinds of compositions as carried out in the work of OHSATO/SEKIGUCHI (1983, 1985) and discussed in section 3.6.

In all these cases extremal solutions of systems are – given the solvability of the system under consideration – available from corresponding extremal solutions of the single equations. Additionally, in interesting special cases of the non-solvability of such systems, these constructions give the best possible approximate solutions – and in general at least nearly the best possible ones. Hence, a "brute-force" methodology can be taken into account which starts by considering such a suitable combination of solutions to the single equations of the system as an a priori approximate solution of the whole system. If doing so, it becomes essential to discuss the quality of an approximate solution. Such a discussion can be provided by using global, but also by using local solvability (or rather, solution) indices. Such indices can also be used to define analogues to confidence intervals (with respect to each point of the corresponding universe of discourse) for the actual approximate solution. Such an approach provides means for evaluating the quality of the actual approximate solution at hand, of the fuzzy model, and hence also for ways to look for improved models.

Therefore we now first reconsider the problem of the comparison of two fuzzy sets. Let C, D be fuzzy sets defined over the same universe of discourse $\mathcal{X}$, i.e. $C, D : \mathcal{X} \to [0,1]$. The following question here is the essential one: to what extent C, D are the same fuzzy set? Or to put it into other words: how distinct are the fuzzy sets C, D?

At the beginning let us recall some main fashions of comparison that have previously been discussed.

(A) The most direct approach is to measure some kind of distance of the membership functions of C and D, i.e. to calculate a distance value $d(C, D)$, cf. KAUFMANN (1975) as well as our distances ϱ_t of definition 3.5. In general, some Minkowski-type distance is quite often employed:

$$d(C,D) = \left(\int_{\mathcal{X}} |\mu_C(x) - \mu_D(x)|^p \, dx \right)^{1/p}, \quad p \geq 1,$$

where we assume the existence of all integrals appearing (which may be guaranteed by restrictions on the types of the membership functions) or have, for finite universes of discourse $\mathcal{X}$, to read them as sums.

As special cases for $p = 1$ one has the Hamming distance, and $p = 2$ yields the usual Euclidean distance.

Other distances which seem to be equally well suited for fuzzy sets are Čebyšev-type distances of the form

$$d(C,D) = \left(\sup_{\mathcal{X}} |\mu_C(x) - \mu_D(x)|^p \right)^{1/p}, \quad p \geq 1,$$

with the well-known Čebyšev metric as the special case $p = 1$.

If the specified distance is equal to zero, we may consider C, D as indistinguishable but, of course, for another choice of a distance function this distance may differ from zero. Note that this way of expressing equality comes almost directly from the well-known concepts of mathematical analysis, and hence does not produce any difficulties in understanding its mathematical background. On the other hand, from the applicational point of view the choice of a proper form of this distance is not obvious, and this requires deeper investigation – usually not of a purely mathematical character.

(B) A second way of comparison of two fuzzy sets is closely related to the so-called possibility measure proposed by ZADEH (1978). Two indices have been put into discussion:

– *possibility* of C with respect to D, $\mathrm{Poss}(C|D)$, given by

$$\begin{aligned} \mathrm{Poss}(C|D) &= \bigvee_{x\in\mathcal{X}} (\mu_C(x)\, \boldsymbol{t}\, \mu_D(x)) \\ &= [\![\exists x(x \,\varepsilon\, C \wedge_{\boldsymbol{t}} x \,\varepsilon\, D)]\!] = \mathrm{hgt}\,(C \cap_{\boldsymbol{t}} D), \end{aligned} \tag{5.1}$$

– *certainty* of C with respect to D, $\mathrm{Cert}(C|D)$, given by

$$\begin{aligned} \mathrm{Cert}(C|D) &= \bigwedge_{x\in\mathcal{X}} (\mu_C(x)\, \boldsymbol{s_t}\, (1 - \mu_D(x))) \\ &= [\![\forall x(\neg x \,\varepsilon\, D \vee_{\boldsymbol{t}} x \,\varepsilon\, C)]\!] \end{aligned} \tag{5.2}$$

with $\boldsymbol{t}$ any t-norm, $\boldsymbol{s_t}$ the corresponding t-conorm, $\bigvee$ for supremum, and $\bigwedge$ for infimum. As a side remark let us mention that in the original formulation ZADEH specified the t-norm as $\boldsymbol{t} = \min$, and the t-conorm $\boldsymbol{s_t}$ as maximum.

The possibility of C with respect to D expresses a degree of overlapping of C and D, whereas $\mathrm{Cert}(C|D)$ is connected with measuring a degree of containment of D in C; but $\mathrm{Cert}(C|D)$ is not the inclusion degree $[\![D \subseteq_{\boldsymbol{t}} C]\!]$ because in (5.2) instead of the implication $p \rightarrow_{\boldsymbol{t}} q$, crucial for defining the fuzzified inclusion $\subseteq_{\boldsymbol{t}}$, a (classically and also for $\boldsymbol{t} = \& = \boldsymbol{t}_L$ – but not in general – equivalent) combination $\neg p \vee_{\boldsymbol{t}} q$ is used.

Remembering that t-norms and their t-conorms are conjugated by the formula (1.9), one obtains

$$\mathrm{Cert}(C|D) = 1 - \mathrm{Poss}(\complement C|D) = 1 - \mathrm{hgt}\,(\complement C \cap D).$$

Now the equality of two fuzzy sets C and D may be evaluated by taking the minimum of degrees $\mathrm{Cert}(D|C)$ and $\mathrm{Cert}(C|D)$; this index of comparison has been discussed in correspondence to some aspects of matching of fuzzy data.

(C) A third approach for the comparison of fuzzy sets uses our previous more set theoretically oriented tools, especially the fuzzified identity $\equiv_t$ of definition 2.3. Modelling the involved connectives for implication and conjunction in our fuzzy set theoretic setting via any residuation operator, i.e. any Φ-operator φ for implication and any (left) continuous t-norm t for conjunction, in addition to our previous equality degree $[\![C \equiv_t D]\!]$ also some *local* degree to which two fuzzy sets C and D are equal each other *at a point* $a \in \mathcal{X}$ is given as a number $\|C = D\|(a)$ defined e.g. in analogy with our definition 2.3 of $\equiv_t$, but deleting the universal quantifier $\forall$ there, i.e. the inf-operator. That means we take this local degree of equality as the value

$$\|C = D\|(a) =_{def} (\mu_C(a) \varphi_t \mu_D(a)) \, t \, (\mu_D(a) \varphi_t \mu_C(a)) \tag{5.3}$$

which is nothing other then the truth degree

$$\|C = D\|(a) = [\![a \varepsilon C \leftrightarrow_t a \varepsilon D]\!].$$

The simple fact that always $\mu_C(a) \leq \mu_D(a)$ or $\mu_D(a) \leq \mu_C(a)$ is the case, i.e. that always $\mu_C(a) \varphi_t \mu_D(a) = 1$ or $\mu_D(a) \varphi_t \mu_C(a) = 1$, allows to simplify (5.3) to

$$\|C = D\|(a) = (\mu_C(a) \varphi_t \mu_D(a)) \wedge (\mu_D(a) \varphi_t \mu_C(a)). \tag{5.4}$$

Thus one has, together with the Φ-operator φ_t, only to take the min-operator instead of the t-norm t to finally find $\|C = D\|(a)$.

Going one step further, nevertheless one usually also likes to have a number expressing a (unique) degree to which C and D are equal to each other in a global sense. For this one has to aggregate the partial evaluations (5.3) and (5.4) of equality over the whole space $\mathcal{X}$.

There is, however, no unique way to perform these tasks. Perhaps the most preferred ways in the engineering community are to take either a so-called optimistic or a pessimistic form of aggregation.

In the *optimistic* case one prefers to modelize the degree of the statement "C and D are equal to each other" by the maximal value of $\|C = D\|(x)$ over $\mathcal{X}$, viz.

$$\|C = D\| =_{def} \sup_{x \in \mathcal{X}} \|C = D\|(x). \tag{5.5}$$

Note that the sup-operator here corresponds to the existential quantifier $\exists$; so one essentially takes only one selected element in $\mathcal{X}$ in which C and D are locally equal at the highest degree. It seems reasonable to look at this "optimistic" global degree as indicating some degree of possibility that C, D may be equal.

On the opposite, i.e. *pessimistic* pole, one often uses the formula

$$\|C = D\| =_{def} \inf_{x \in \mathcal{X}} \|C = D\|(x). \tag{5.6}$$

This degree of equality has a "pessimistic character" because obviously this degree (5.6) indicates the worst case of all the local degrees (5.3). Here as always the inf-operator corresponds to the universal quantifier $\forall$ and therefore (5.6) can be rewritten as

$$\|C = D\| = [\![\forall x(x \,\varepsilon\, C \leftrightarrow_t x \,\varepsilon\, D)]\!].$$

Therefore according to proposition 1.26 (i) this means

$$\begin{aligned} \|C = D\| &= [\![\forall x((x \,\varepsilon\, C \rightarrow_t x \,\varepsilon\, D) \wedge_t (x \,\varepsilon\, D \rightarrow_t x \,\varepsilon\, C))]\!] \\ &\geq [\![\forall x(x \,\varepsilon\, C \rightarrow_t x \,\varepsilon\, D) \wedge_t \forall x(x \,\varepsilon\, D \rightarrow_t x \,\varepsilon\, C))]\!] \\ &= [\![C \subseteqq_t D \wedge_t D \subseteqq_t C]\!] = [\![C \equiv_t D]\!] \end{aligned}$$

and thus is closely analogous to our previous equality degree.

Generalized quantifiers in the sense of the standard model theory of mathematical logic like "almost everywhere" instead of these classical "sometimes" or "always" versions, or even fuzzy quantifiers like "most", "not too few" applied to the localized equality values of C, D would give global evaluations in between the optimistic and the pessimistic point of view. Another possibility to get an intermediate value of global equality not as likely to cause overestimation or understimation as (5.5) or (5.6) is to take for example the average value of all the local equality degrees, i.e. to consider

$$\|C = D\| =_{def} \sum_{x \in \mathcal{X}} \|C = D\|(x) \,/\, \mathrm{card}\,(\mathcal{X}) \tag{5.7}$$

in the case of a finite universe of discourse $\mathcal{X}$ or of at least a finite support of the fuzzy set $C \cup D$, and to consider correspondingly the normalized integral, assuming P-integrability of the function $\|C = D\| : \mathcal{X} \to \mathbb{R}$,

$$\|C = D\| =_{def} \frac{1}{\mathrm{card}\,(X^{[1]})} \int_{\mathcal{X}} \|C = D\|(x)\, dP$$

in the case of a continuous, finitely P-measurable universe of discourse $\mathcal{X}$ or of at least a finitely P-measurable support supp$(C \cup D)$.

In the sequel we will also utilize some characteristics that can be deduced from histograms of the values of the local equality degrees (5.3) computed for each element of the universe of discourse.

An interesting problem that arises in relation to such a local equality degree (5.3) is related to the so-called *inverse problem*, cf. PEDRYCZ (1990a). In concise formulation we can state it as follows:

> for a given fuzzy set C and a certain value v determine such a fuzzy set D for which the inequality
>
> $$\|C = D\|(x) \geq v \tag{5.8}$$
>
> holds true for all points x of the universe of discourse.

In other words, one is interested in getting a fuzzy set D such that C and D are equal to each other to a degree locally not lower than v. (The equality of C and D again here is considered in the pointwise sense).

Omitting a detailed discussion which has been performed in PEDRYCZ (1990a), it is worth recalling only a few main results. At the very beginning notice that (5.8) is usually not uniquely solvable. Instead, one can indicate a certain range of the unit interval in which the values of the membership function of D may be taken to fulfill (5.8). Thus, the resulting fuzzy set D can in fact be interpreted as an interval-valued fuzzy set, a so-called Φ-fuzzy set, cf. SAMBUC (1975).

It is remarkable to note that replacing the inequality sign in (5.8) by equality may cause there to be no solution. Thus, our formula (5.8) forms an appropriate problem statement.

A few facts now will be summarized to give properties of the fuzzy sets forming a solution to (5.8). We start by rewriting (5.3) in a slightly modified form utilizing properties of the Φ-operator $\varphi_{\boldsymbol{t}}$, which starting from a continuous t-norm $\boldsymbol{t}$ is given by (1.19). Straightforward calculations yield

$$\|C = D\|(x) = \begin{cases} \mu_C(x)\,\varphi_{\boldsymbol{t}}\,\mu_D(x), & \text{if } \mu_D(x) < \mu_C(x) \\ \mu_D(x)\,\varphi_{\boldsymbol{t}}\,\mu_C(x), & \text{if } \mu_D(x) > \mu_C(x) \\ 1, & \text{if } \mu_D(x) = \mu_C(x). \end{cases} \tag{5.9}$$

Let us denote by $D_*^v(a)$ the set of values of the membership function $\mu_D(x)$ at point a satisfying (5.8), i.e.

$$D_*^v(a) =_{def} \{\mu_D(a) \in [0,1] \mid v \leq \|C = D\|(a)\}.$$

Thus, if we put $v = 0$ then obviously $\mu_D(a) \in [0,1]$ without restriction, i.e. μ_D may take any values between 0 and 1; therefore $D_*^v(a) = [0,1]$. On the opposite side, putting $v = 1$ reduces $D_*^v(a)$ to exactly one point: $D_*^v(a) = \{\mu_C(a)\}$. For intermediate situations, viz. for $v \in (0,1)$, the set $D_*^v(a)$ is derived by solving the two inequalities

$$\mu_C(a)\,\varphi_t\,\mu_D(a) \geq v, \tag{5.10}$$

$$\mu_D(a)\,\varphi_t\,\mu_C(a) \geq v. \tag{5.11}$$

Solving the first of them we get a subinterval $[d_1, \mu_C(a)]$ while the second leads us to a subinterval $[\mu_C(a), d_2]$ where d_1 and d_2 are determined by solving (5.10) and (5.11) respectively. Finally, $D_*^v(a)$ is nothing other than the union of both those intervals:

$$D_*^v(a) = [d_1, \mu_C(a)] \cup [\mu_C(a), d_2] = [d_1, d_2]. \tag{5.12}$$

Definition 5.1 *The set (5.12) of solutions $D_*^v(a)$ will be called* equality interval *or* confidence interval *for the fuzzy set D (at point a) induced by v for the given fuzzy set C.*

This inverse construction seems to be of considerable use in determining resulting fuzzy sets for fuzzy models constructed in the framework of fuzzy relational equations. In order to produce a more detailed picture of what goes on in this area and what form of solution can be obtained there, we reconsider this topic in the last part of section 5.4.

5.2 Approximate solutions for systems of fuzzy relational equations

As it becomes quite clear from the previous discussions the results for example of Table 3.2 have a significant value for applications only in the case that solutions really exist, i.e. that not only approximate solutions exist. If this true solvability is not the case – and it is this more uncomfortable situation one usually meets in practice[1] – then one has to think about other ways to overcome the problem of the nonexistence of (true) solutions.

[1] The situation here with the fuzzy data set $(A_i, B_i), i = 1, \ldots, n$, is quite analogous to the usual situation one has with nonfuzzy data e.g. from a sequence of measurements: theoretically, perhaps, they should be points on some curve, but in practice this only approximately happens to be the case.

A quite simple, and perhaps for the practitioner the most obvious, way out is to use the formulas which describe solutions – in the case of solvability – of systems of fuzzy equations like

$$\Theta(R, A_i) = B_i, \qquad i = 1, \dots, n$$

even if there does not exist a solution to the system of fuzzy relational equations to be considered – and then to check the quality of the approximate solution obtained in this way. If this quality is not too bad, i.e. if results/outputs $\overline{B}_i$ given for the corresponding data/inputs A_i by the approximate solution are not "too far" from the results/outputs B_i one intends to get from the set of data (A_i, B_i) that constitutes the intended model and hence determines the considered system of fuzzy relational equations, in such a case at least an approximate fuzzy model is realized and can be used within some quality bounds.

Taking such a path is completely in accordance with fuzzy modelling, because any fuzzy model aims at a rough, i.e. approximate description of real processes, situations, and the like.

Recapitulating the central results of chapter 3, which are of course central from the present point of view, the point already made in section 3.5 was that we found best possible approximate solutions for single equations, cf. theorems 3.4, 3.7, and also found suitable candidates for "nearly best possible" approximate solutions to systems of fuzzy equations in theorem 3.9. Section 3.6 added, with the results mentioned there in Facts 1 to Fact 4, cf. page 131f., some further generalizations for more abstract types of fuzzy equations.

These results are all together now the background for some kinds of reasonable guesses we have in mind concerning the determination of approximate solutions.

Claim 1: For systems of fuzzy relational equations of sup-type consider the intersection of all the greatest solutions of the single equations (or of good approximations for them) as a suitable approximate solution for the system.

Claim 2: For systems of fuzzy relational equations of inf-type consider the union of all the smallest solutions of the single equations (or of good approximations for them) as a suitable approximate solution to the system.

Claim 3: And for "combined" types of equations consider a suitable "combined strategy" to provide a reasonable guess for an approximate solution.

Applying this kind of "brute-force" strategy, in every case one needs some accompanying discussion of the (expected) quality of such an approximate solution. Of course, besides this "brute-force" approach towards solving systems of fuzzy relational equations there are other, competing or supplementary ways to treat this problem of approximately "solving" such systems.

Bearing in mind that in general no solution will exist, it is of special interest to discuss ways of making any such system of fuzzy relational equations "better solvable" up to a level of approximate solvability acceptable by the intended model-builder or user. In general, two ways of approach may be distinguished here. The first is closely related to such changes in the data set that might make a set of equations "more" solvable in the sense of the existence of better approximate solutions. The second is oriented toward the modification of the structure of the equations, for example by taking another form of equation, or simply by extending the structure of the equations, e.g. by adding some new variables (often called *explanatory* ones). We list some representative methods coming from these two groups and refer to the literature where more detailed descriptions may be found. We also recall some underlying ideas supporting these methods.

While considering the first possible stream of modification we have to pay attention to the fact that again at least two main groups might be recognized: in the first we simply skip some elements of the data set (in fact this deletion modifies it), whereas in the second all the elements of the data set are modified.

1. In GOTTWALD/PEDRYCZ (1985) a way of measuring the consistency of pairs of the data set, say (A_i, B_i) and (A_j, B_j), has been proposed. Roughly speaking, the procedure works as follows: A suitable consistency index has to be considered that takes into account for pairs $(A_{i_0} \models\!\!\Rightarrow B_{i_0})$ and $(A_{j_0} \models\!\!\Rightarrow B_{j_0})$ of control rules an absolute difference between the possibility of A_{i_0} versus A_{j_0} and the possibility of B_{i_0} versus B_{j_0}. If for the two specified indices i_0 and j_0 a low value is obtained (e.g. almost equal to zero) then one pair of this data set is expected to be a candidate for being removed from the entire data set.

The simple idea behind this approach is that it is taken as reasonable to suppose that "nearly equal" input data should not be combined with "quite different" output data.

A further variant of such a type of approach may be first to split the data set into some clusters of control rules which within each cluster are similar to each other. Then either

a simpler set of control rules can be taken into account with some "centres" of those clusters – or each cluster can be transformed into one rule, but with probabilistic sets as input and output values.

2. A different approach was proposed in GOTTWALD/PEDRYCZ (1986) and reconsidered in the previous chapter, section 4.3. There all the fuzzy sets $A_i \in \mathbb{F}\mathcal{X}$ are modified by replacing them by their modified version:

$$A_i \;\mapsto\; A_i^\alpha = A_i \cup X^{[\alpha]}$$

where α is a threshold level supposed to be provided by an inspection of the concrete applicational situation. The basic idea behind this data transformation (4.63) is that the smallest values of the membership functions of A_i are irrelevant: for example they may be very sensitive to existing noise and therefore they should be masked by an appropriately chosen value of the threshold level α. As proved in that section, the solvability degree discussed there is an increasing function of α. Then by observing its dependency of the threshold α, one can take a suitable value of α which on one hand is low enough not to disturb the data set at a significant level, and on the other hand can lead to a significant increment of the solvability degree.

3. Yet another approach follows again a different way. Here the fuzzy sets A_i are replaced by "sharpened" versions; more precisely, instead of the original (input) data A_i, $i = 1, \ldots, n$, one takes pairwise disjoint families of fuzzy sets $(A_i')_{1 \leq i \leq n}$, i.e. such which fulfill the condition

$$A_i' \cap_t A_j' = \emptyset \quad \text{for all } 1 \leq i < j \leq n. \tag{5.13}$$

In this case, if for each pair (A_1', B_i) there exists a solution to the corresponding equation, then one is sure that there exists a solution for the entire system of equations. The background facts here are the results of section 4.2 which indicate that the pairwise disjointness of the "input" data $(A_i')_{1 \leq i \leq n}$ is always a sufficient condition for the non-interactivity of that system (seen as a realization of a system of control rules). And non-interactivity of course guarantees solvability.

4. A proposal formulated by WAGENKNECHT/HARTMANN (1986, 1986a) deals with modifications of the fuzzy sets B_i forming the right-hand parts of the equations. Instead of using the fuzzy sets B_i, so-called fuzzy sets with tolerances have been utilized. By a fuzzy set with a tolerance they essentially mean an interval-valued fuzzy set. At present, the tolerances are attached in a heuristic fashion to reach a situation where the entire set of equations has a solution. Nevertheless, the tolerances should be adjusted by a user and then there is no security that they are not taken too broadly or that their choice does not guarantee the existence of a solution.

5. Yet another way has been discussed in LI (1985/86), where a certain construction has been designed to perturbate the data set and to reach a state where the set of equations has a solution.

6. In HIROTA/PEDRYCZ (1983) the use of the concept of probabilistic set has been discussed. In this application, a probabilistic set is used to mean a certain class of fuzzy sets that are similar to each other. As a consequence, the data set is split into classes, i.e. it is structured. Afterwards it is attempted to solve a reduced set of equations just by taking representatives of each class, i.e. prototypes for each class.

7. Instead of measuring either the consistency of data pairs (A_{i_0}, B_{i_0}), (A_{j_0}, B_{j_0}) as mentioned in point **1.** or discussing the degrees of equality of the intended and the realized output (for some given input) as we did in earlier chapters, as well as in section 5.1, one could start from the data set of pairs (A_i, B_i), $i = 1, \ldots, n$, together with some given distance d for fuzzy subsets of the output space.

Now, looking in the traditional mathematical style for an optimization problem, one can pose and discuss the problem to find such a fuzzy relation R, i.e. to find a $(\mathrm{card}(\mathcal{X}) \cdot \mathrm{card}(\mathcal{Y}))$-matrix R, for which

$$\sum_{i=1}^{n} d(B_i, R''A_i) = \sum_{i=1}^{n} d(B_i, A_i \circ_t R) \;\Rightarrow\; \min!$$

under the constraints that

$$\mu_R(a,b) \in [0,1] \qquad \text{for all } a \in \mathcal{X}, b \in \mathcal{Y}.$$

This is a problem with $\mathrm{card}(\mathcal{X}) \cdot \mathrm{card}(\mathcal{Y})$ variables and thus will rather quickly present dimensionality problems because in interesting applications the input and output spaces $\mathrm{card}(\mathcal{X})$, $\mathrm{card}(\mathcal{Y})$ will be finite sets which are not too big but which also are not very small finite sets.

Therefore, perhaps, another way round could prove to be more efficient: given such a distance function d, try to express it (at least approximately) as a distance function ϱ_t and then apply the "logical" methods of chapter 3.

Recapitulating this first stream of methods we can observe that they modify the data set to make the corresponding system of equations "more solvable" just by discovering a structure in it or in one of its "slightly" modified forms – modified for example by the elimination of "outliers" in the data set.

Discussing the second main stream of investigation, it is worth recalling some ideas which are of interest for it.

A) First, keeping in mind that now the structure of the equation is supposed to be fixed (viz. the type of composition) and moreover the fuzzy sets A_i are unchanged, one can make some parametric studies either for modifying some t-norm or Φ-operator appearing in the equation, or look for a suitable value of the parameter in any parametrized class of, for example, t-norms. This may lead to higher accordance of the data provided and the fuzzy sets generated by the equations. Unfortunately, till now almost any method for a systematic search within the whole infinite set of t-norms or t-conorms has been lacking. Some partial search is nevertheless possible if one restricts the considerations, for example, to one of the parametrized families of t-norms mentioned in Table 1.1.

B) Contrary to the previous approach which might be seen as a parametric adjustment, one is now interested in considering different structures of the fuzzy relational equations. Restricting ourselves to those already listed, we have four ways when fitting the basic types of equations together with the convex combination of OHSATO/SEKIGUCHI (1983), which always has a solution in the case of a suitable choice of the convex combinator λ, a (fuzzy) parameter controlling the influence of the "simple" types of equations which are combined.

C) A third way that allows one to reach the main goal, i.e. to get a solvable system of equations, is to add a new fuzzy variable which "separates" the fuzzy sets $A_1, A_2, \ldots, A_n$, cf. PEDRYCZ (1988, 1990). In its spirit this way is similar to the method of GOTTWALD/PEDRYCZ (1986) but now no input fuzzy set $A_1, \ldots, A_n$ is modified. To give a straightforward explanation just take a system of equations

$$A_i \circ_t R = B_i, \quad i = 1, \ldots, n. \tag{5.14}$$

If $\boldsymbol{R}_i \neq \emptyset$ for each of the solution sets $\boldsymbol{R}_i$ of the i-th equation, then to make the system solvable it is enough to consider a new universe of discourse $\mathcal{Z} = \{z_1, \ldots, z_n\}$ and to define fuzzy singletons $Z_1, \ldots, Z_n$ as $Z_i = \langle\langle z_i \rangle\rangle_1$, i.e. by

$$\mu_{Z_i}(z_j) =_{def} \begin{cases} 1, & \text{if } i = j \\ 0 & \text{otherwise.} \end{cases} \tag{5.15}$$

Then this new fuzzy variable defined in $\mathcal{Z}$ allows one to be sure that the system of equations

$$(A_i \times Z_i) \circ_t R = B_i, \quad i = 1, \ldots, n \tag{5.16}$$

has a solution. Note, however, that now R is defined in the cartesian product of the universes $\mathcal{X}, \mathcal{Y}, \mathcal{Z}$. In other words, the enriched structure (5.16) of (5.14) is sufficient to have the system of equations solvable. The new universe of discourse $\mathcal{Z}$ used here is a "space of explanation" and obviously in close analogy to additional variables discussed for example in statistical models.

The weakest point of this approach is perhaps the special choice of the separating fuzzy sets Z_i chosen as fuzzy singletons – which therefore have no linguistic meaning, i.e. usually may not reflect any linguistic concept connected with the process whose control is intended to be realized using the control rules $A_i \models\!\!\Rightarrow B_i$. Nevertheless, one can search for a way to fuzzify these singletons by just remembering not to exceed a certain overlap, measured for example as the height of the intersections $Z_i \cap Z_j$, between these auxiliary fuzzy sets. For a more formal discussion see PEDRYCZ (1988).

Until now we have presented ideas how a degree of solvability of a given system of equations can be enlarged; we have not discussed how the property of (partial) lack of solvability influences the precision of the results obtained by means of the fuzzy model constructed in such a way. This forms the topic of section 5.4.

5.3 Fuzzy equations for processing fuzzy data

Bearing in mind the results obtained in section 4.3 with regard to data transformations, let us take a look at their role in the processing of fuzzy data. The problem may actually be formulated in the following way.

Some real world system shall be given that has to be modelled. Furthermore it is assumed that there are some fuzzy pieces of evidence associated with this system which may be described by a collection of conditional statements:

$$\textsf{if } A_i \textsf{ then } B_i, \qquad i = 1, \ldots, n.$$

The notion of system is considered here in a broad sense of the word. Thus a system may be treated as a dynamic system with an unknown structure as in DUBOIS/PRADE (1980), PEDRYCZ (1982), ZADEH (1971a) or as a global knowledge base (explanatory data base) for the manipulation of commonsense knowledge as in ZADEH (1983, 1984). The given data A_i, B_i are supposed to be linguistic terms, i.e. linguistic values of suitable (input and output) variables, describing ambiguous facts available at hand. These linguistic terms are expressed as fuzzy subsets $A_i : \mathcal{X} \rightarrow [0,1]$ and $B_i : \mathcal{Y} \rightarrow [0,1]$ of suitable universes of discourse.

In order to process fuzzy data one has to build a model of the system for which the above-mentioned data are relevant. That means that one has a "reality layer" and a "modelling layer" which have to correspond to each other sufficiently well in the relevant aspects of the problem.

In the modelling layer a surface model structure combining the fuzzy data A_i and B_i in the sense of the fuzzy relational equations (4.28) is assumed. The fuzzy relation R of the model results from a rule-based description of the system behaviour from which the fuzzy data A_i, B_i is available. Of course, not all the data pairs (A_i, B_i) for $i = 1, \ldots, n$ may satisfy the system of equations which fixes the model as supposed above. This is due to several reasons:

- the structure of the model does not perfectly follow the structure of the system which is indeed fuzzy; for instance, we have no extra information on logical connectives – viewed as the concrete t-norm – of the model;

- the data standing at our disposal are not unbiased – they may perhaps be corrupted by some noise. (Note that the problem of their true characterization is not as evident as in signal analysis performed in for example communication theory for signals with additive Gaussian noise.)

The approach we follow here is in modifying the data by imposing some threshold level. The idea is that lower values of the membership functions of A_i and B_i are less reliable than the highest ones. Thus, the threshold level α is used to convert the original data set, i.e. the pairs (A_i, B_i) for all $i = 1, \ldots, n$ into a modified one (A_i^α, B_i^α), $i = 1, \ldots, n$ according to the data transformation (4.63).

The fuzzy relation of the model is then calculated according to theorem 3.9 (ii), i.e. as

$$R = \bigcap_{i=1}^{n} (A_i^\alpha \triangleright_t B_i^\alpha)$$

while the solvability degree $\xi(\alpha)$ of the system of equations is chosen according to (3.12) but derived from the system of fuzzy relation equations modified according to the manipulation procedure (4.63).

Any discussion of the value α of threshold level should embrace two facts:

(i) higher values of α allow us to achieve higher values of the solvability index $\xi(\alpha)$;

(ii) higher values of α and hence more extended modifications of the original data set lead to more rough models of the original process and less transparent model relations.

Fact (i) has been proved in previous sections. Fact (ii) comes from the property of Φ-operators that the values of the membership function of R, calculated according to (5.3), are not smaller than α. It implies in turn that the structure of the relation of the original model becomes partly hidden. Of course, for $\alpha = 1$ there is no remaining structure – the membership function of R becomes simply equal to $(X^{[1]} \times Y^{[1]})$ and hence corresponds to the linguistic term "unknown". If one agrees to accept the lower values of threshold α, a certain structure of the model appears but its adequacy (measured via the solvability degree) decreases. All in all, increasing values of the threshold cause improvements in the internal adequacy of the model R in the sense of its behaviour in accordance with the constituting rules, but the ability of recognition of the structure of the fuzzy model falls.

Some illustration of the behaviour of $\xi(\alpha)$ is shown in Figure 5.1.

Case 5.1a illustrates an ideal situation where the entire collection of the data perfectly fits the equation of the model. A situation lying almost on the opposite pole is shown in Figure 5.1b. Now a slight change of the threshold α down the value 1.0 gives the value zero for the solvability degree. Case 5.1c represents an intermediate situation where the structure of the model may

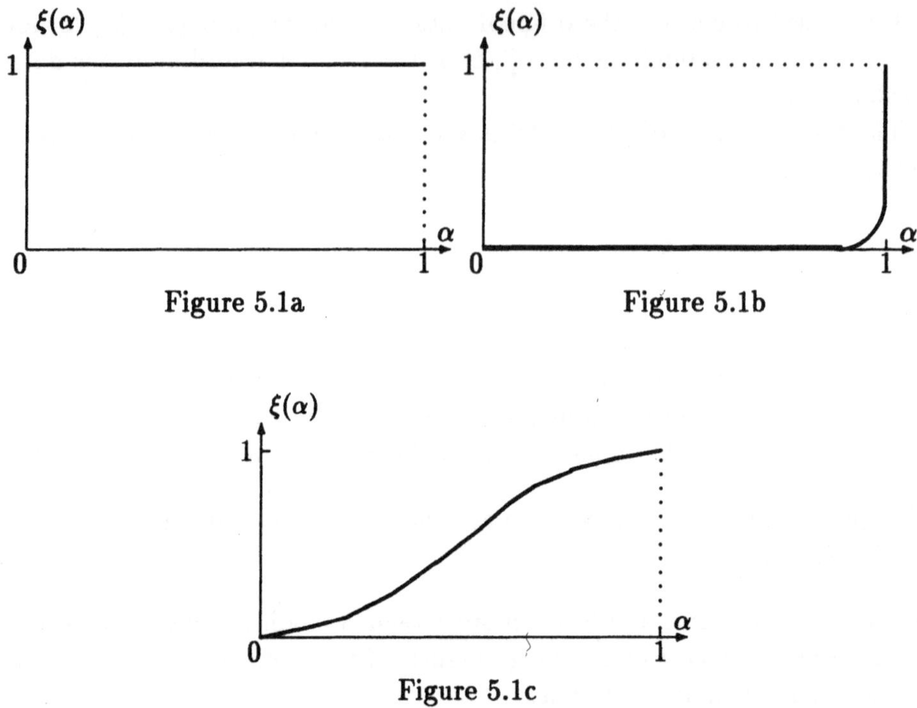

Figure 5.1: Examples of solvability degrees $\xi(\alpha)$ as functions of the threshold level α

be discovered to a certain level without drastic reduction in the value of the solvability degree.

In such a sense $\xi(\alpha)$ is interpreted as a quality index of the fuzzy model that is going to be constructed. The model characterized by the solvability degree of Figure 5.1a is perfect, while for Figure 5.1b it is (almost) meaningless. This remark opens a new way of comparison of two fuzzy models: we say that a first one M_1 is *superior* to a second one M_2, $M_1 \succ M_2$, iff there holds

$$\xi^1(\alpha) \geq \xi^2(\alpha) \quad \text{for all} \quad \alpha \in [0,1],$$

where the superscript attached to the solvability degree refers to the model. Now, for example, one can search for an appropriate t-norm to be used in the model equations which leads to a high value of $\xi(\alpha)$.

Let us consider a certain policy that leads to some "slightest" modification of all the fuzzy data pairs (A_i, B_i), $i = 1, \dots, n$ and which is sufficient to achieve the value =1 for the solvability degree. Denote by α_0 the smallest value of the threshold level for which the solvability degree equals one:

$$\alpha_0 = \min\{\alpha \in [0,1] \mid \xi(\alpha) = 1\}.$$

Solve the system of fuzzy relational equations

$$A_i^{\alpha_0} \circ_t R = B_i^{\alpha_0}, \quad i = 1, \dots, n$$

with respect to the fuzzy relation R. This yields the fuzzy relation R^* given as

$$R^* = \bigcap_{i=1}^{n} (A_i^{\alpha_0} \triangleright_t B_i^{\alpha_0}). \tag{5.17}$$

Now convert all the fuzzy data B_i to B_i^* according to the following formula:

$$B_i^* = A_i \circ_t R^*, \quad i = 1, \dots, n.$$

This in turn gives a modified set of pairs of fuzzy data (A_i, B_i^*), $i = 1, \dots, n$ whose corresponding system of model equations has value =1 of its solvability degree.

Being equipped with the fuzzy model in the form of a system of fuzzy relational equations or the fuzzy relation R determined by such a system and its evaluation by $\xi(\alpha)$, the determination of the fuzzy consequence, i.e. the fuzzy output B for a given fuzzy input A may then be performed in the following steps:

- evaluation of the model quality via its expected outputs and the solvability degree ξ,

- modification of the result of this composition according to (4.63) with respect to the threshold level α_0, where α_0 may be the lowest value of the threshold level that gives an accetable value of the relative solvability degree $\xi(\alpha)$,

- determination of the corresponding "approximate model" R^*,

- calculation of the composition of the input A with the "approximate" fuzzy relation R^*.

It is remarkable that in a situation in which α_0 is high enough, a fuzzy set of the form B^{α_0} becomes nearly meaningless (too "fuzzy"). Moreover, one cannot modify the result of the composition of A and R^*, but in such a case it should be underlined that the values of the membership function which are below α_0 should be neglected, and we are unable to treat them as significant.

The effect of deformation of B into B^{α_0} may additionally be measured by some energy measure G of the fuzzy set ΔB^{α_0} defined as

$$\mu_{\Delta B^{\alpha_0}}(y) =_{def} \mu_{B^{\alpha_0}}(y) - \mu_B(y) \quad \text{for all } y \in \mathcal{Y}$$

preferably some energy measure G with $G(\emptyset) = 0$ but whose values are $\neq 0$ for fuzzy singletons.

5.4 Evaluating the quality of fuzzy models

In the previous sections we have summarized among other things how fuzzy relational equations can be solved in an exact manner, or at least in an approximate fashion. Now we intend to move one more step forward and to discuss further ways in which the precision of a solution of fuzzy relation equations and also of other fuzzy models for given sets of fuzzy data may be expressed quantitatively. Immediately one concludes that this might have primordial influence on the application of constructed fuzzy relational equations. In the case of a broad utilization of such fuzzy models one likes to have at hand a clear answer what quality of results can be expected if the fuzzy model has been chosen – and also how to compare different fuzzy models for the same set of fuzzy data or the same process.

The most essential and fruitful idea such evaluations and comparisons shall be based upon is quite often some kind of "performance index" which is (mainly) intended to be used for relative evaluations of fuzzy models.

5.4.1 Evaluating fuzzy models through fuzzy integrals

The "localization" (5.3) of the generalized equality considerations we mentioned in section 5.1 shall now also be used in another way.

Suppose for simplicity that the universe of discourse $\mathcal{Y}$ of the output fuzzy sets B_i of some fuzzy controller with a family (4.1) of control rules is a finite set: $\mathcal{Y} = \{b_1, \ldots, b_m\}$, and consider fuzzy sets $C, D \in I\!F(\mathcal{Y})$. Let be for each index $j = 1, \ldots, m$:

$$\gamma_j^{C,D} =_{def} \|C = D\|(b_j)$$

according to (5.3) and (5.4). That means with $\gamma_j^{C,D}$ we "measure" pointwise the coincidence of the membership degrees of C and D getting for example $\gamma_j^{C,D} = 1$ in the case that $\mu_C(b_j) = \mu_D(b_j)$.

Suppose additionally that there has been determined some fuzzy relation $\overline{R}$ as a – perhaps only approximate – realization of the fuzzy controller constituted by (4.1). Because $\overline{R}$ may only approximately realize the control rules (4.1) let additionally be

$$\overline{B}_i = \overline{R}'' A_i = A_i \circ_{\boldsymbol{t}} \overline{R}$$

the real output of $\overline{R}$ for the input A_i. To simplify notation put

$$\gamma_j^i =_{def} \gamma_j^{B_i,\overline{B}_i} = \|B_i = \overline{B}_i\|(b_j). \tag{5.18}$$

Now combine those indices for all control rules into

$$\Gamma_j =_{def} \bigwedge_{i=1}^{n} \gamma_j^i \tag{5.19}$$

and unite all these local comparison indices into the *evaluation vector*

$$\Gamma =_{def} (\Gamma_1, \ldots, \Gamma_m). \tag{5.20}$$

Instead of looking at Γ as a vector of local comparison indices, one can also view Γ as the fuzzy subset of $\mathcal{Y}$ of all those points where the fuzzy model fits in well with the fuzzy data set.

The essential idea now is to use this vector or fuzzy set Γ to evaluate the global property

Fuzzy model $\overline{R}$ well represents the fuzzy data set over the whole output space $\mathcal{Y}$. (5.21)

The reference to formula (5.4) in the definition of the present indices γ_i^j has as an immediate consequence an inequality for the indices Γ_j: an inequality which relates to the dependence of those indices on the t-norm which defines the Φ-operator which is involved in formula (5.18), i.e. in formula (5.4). Writing for the moment $\Gamma_j^{(\boldsymbol{t})}$ in the case that the definition (5.19) is (implicitely) referring back to the t-norm $\boldsymbol{t}$, we immediately have for any left continuous t-norms $\boldsymbol{t}_1, \boldsymbol{t}_2$ that

$$\boldsymbol{t}_1 \leqq \boldsymbol{t}_2 \;\Rightarrow\; \Gamma_j^{(\boldsymbol{t}_1)} \leq \Gamma_j^{(\boldsymbol{t}_2)}$$

and thus that $\Gamma_j^{(\min)}$ conveys the most pessimistic, i.e. most restrictive evaluation of the (localized) equality of fuzzy sets.

If $\Gamma = \underline{1}$ is the vector with all its components =1 then one may assume that the model $\overline{R}$ of the process under consideration completely fits the data set (4.1). If $\Gamma \neq \underline{1}$ a degree of fitness of the model, i.e. of the fuzzy controller $\overline{R}$ will now be expressed through a fuzzy integral.

Two different fuzzy models R^1, R^2 of the same set of fuzzy data, e.g. the same system (4.1) of control rules, can be compared with respect to property (5.21) in an almost obvious way if their respective evaluation vectors Γ^1, Γ^2 of local equality degrees of realized and expected outputs chosen according to (5.20) are comparable:

If for two fuzzy models R^1, R^2 of the same set of fuzzy data for their evaluation vectors Γ^1, Γ^2 one has $\Gamma^1 \geqq \Gamma^2$, i.e. $\Gamma_j^1 \geq \Gamma_j^2$ for all $j = 1, \ldots, m$, then the fuzzy model R^1 is the better one of the two.

Unfortunately, however, this situation of comparable evaluation vectors Γ^1, Γ^2 is an exception. Usually these vectors are incomparable with respect to their natural componentwise partial ordering. This fact, as well as the pointwise construction of Γ with respect to the points of the output space $\mathcal{Y}$, suggests realizing a partial evaluation of the quality of fuzzy models relative to each point of $\mathcal{Y}$, i.e. considering instead of the global property (5.21) their local version

$$\text{\textit{Fuzzy model } } \overline{R} \text{\textit{ well represents the fuzzy data set at the point } } b_k \text{\textit{ of the output space } } \mathcal{Y}. \tag{5.22}$$

In the spirit of the Gestalt principle, cf. for example CORGE (1975), it is obvious that the evaluation of the global property (5.21) cannot be deduced by simple, perhaps even linear aggregation of the partial evaluations of the model given using (5.22). This leads to the idea of considering some fuzzy measure and the fuzzy integral defined by it as plausible tools for formulating the global evaluation of the fuzzy model out of its local evaluations.

By a *fuzzy measure* G over the universe of discourse $\mathcal{Y}$, as defined in SUGENO (1974, 1977), a real valued function $G : \mathbb{P}(\mathcal{Y}) \to [0,1]$ over the power set $\mathbb{P}(\mathcal{Y})$ of $\mathcal{Y}$ is meant which has the properties

(FM1) $G(\emptyset) = 0 \quad \text{and} \quad G(\mathcal{Y}) = 1,$

(FM2) for all $\mathcal{B}_1, \mathcal{B}_2 \subseteq \mathcal{Y}$

$$\mathcal{B}_1 \subseteq \mathcal{B}_2 \Rightarrow G(\mathcal{B}_1) \leq G(\mathcal{B}_2),$$

(FM3) for all $\subseteq$-monotonic sequences $(\mathcal{B}_n)_{n\geq 1}$ of subsets of $\mathcal{Y}$

$$\lim_{n\to\infty} G(\mathcal{B}_n) = G(\lim_{n\to\infty} \mathcal{B}_n).$$

The second of these properties, the monotonicity of the fuzzy measure G, is of crucial interest here.[2]

The *fuzzy integral* based on such a fuzzy measure G is defined for any[3] function $H : \mathcal{Y} \to [0,1]$, i.e. any fuzzy subset H of $\mathcal{Y}$ and any (crisp) subset $\mathcal{B}$ of $\mathcal{Y}$ by the formula

$$\int_{\mathcal{B}} H(x) \circ G(.) =_{def} \sup_{\alpha\in[0,1]} \min\{\alpha, G(\mathcal{B} \cap H^{\geq\alpha})\}. \tag{5.23}$$

The idea now is that a fuzzy measure G "measures" with its values $G(\mathcal{B})$ to what extent one can judge the quality of a fuzzy model with respect to the global property (5.21) out of the local variants (5.22), i.e. from the knowledge of the localized information Γ_j of (5.19) for points in $\mathcal{B}$. And the fuzzy integral provides this global evaluation from the local ones.

Assuming that one has the relevant information Γ_j on the local behaviour of a fuzzy model for all points b_j of a subset $\mathcal{B}$ of the output space $\mathcal{Y}$ and also the fuzzy measure G available then an evaluation of the fuzzy model, based on that partial information and using the fuzzy integral, can be provided by the index

$$\Lambda(\mathcal{B}, \Gamma) =_{def} \int_{\mathcal{B}} \Gamma(x) \circ G(.) \tag{5.24}$$

if we write $\Gamma(b_j) = \Gamma_j$ in this formula for ease of notation. Of course, the ideal situation for obtaining the global evaluation of the fuzzy model is to have $\mathcal{B} = \mathcal{Y}$. In that case formula (5.23) becomes simpler and the fuzzy model is then evaluated by the index

$$\Lambda_\Gamma =_{def} \int_{\mathcal{Y}} \Gamma(x) \circ G(.) = \sup_{\alpha\in[0,1]} \min\{\alpha, G(\Gamma^{\geq\alpha})\}.$$

[2]This monotonicity property replaces the stronger additivity property of the usual measures. By the way, let us note that the continuity property (FM3) is of course inessential for fuzzy measures over finite sets.

[3]In the case of an infinite universe of discourse $\mathcal{Y}$ the matter becomes a little more difficult. As with traditional measures and integrals, one has to restrict the definition of the measure to some suitable σ-algebra of subsets of $\mathcal{Y}$ and is only able to integrate over functions $H : \mathcal{Y} \to [0,1]$ which are measurable with respect to that σ-algebra.

The monotonicity (FM2) of the fuzzy measure now reflects the fact that the knowledge of more local information always adds to the global evaluation of the fuzzy model. Additionally, however, the fuzzy measure allows the "splitting" of the output space $\mathcal{Y}$ into regions of points whose local evaluations of the fuzzy model are relatively more essential for the global evaluation than points of other regions. That means that one becomes able to distinguish points of the output space which are of "central importance" for the global evaluation of the fuzzy model from such ones which are not so important for the global evaluation. In practical applications the points of central importance may then be such ones of a region in $\mathcal{Y}$ which is crucial for the overall behaviour of the fuzzy model because the model has to act quite exactly there – and other regions which allow for a rougher model behaviour can have a lower significance for the global evaluation.

Nevertheless, the information provided by the index $\Lambda(\mathcal{B}, \Gamma)$ from (5.24) is quite incomplete in the case that $\mathcal{B} \neq \mathcal{Y}$ because one is unable to distinguish the lack of information about the local behaviour of the fuzzy model from the bad quality of this model. Therefore it seems preferable to switch from this index (5.24) to another index Ξ, which itself is an ordered pair consisting of the value $G(\mathcal{B})$ of the "weighted" portion of available information on the one hand, and of the normed fuzzy integral (5.24) on the other hand:

$$\Xi(\mathcal{B}, \Gamma) =_{def} \left(G(\mathcal{B}), \frac{\Lambda(\mathcal{B}, \Gamma)}{G(\mathcal{B})}\right) = \left(G(\mathcal{B}), \frac{1}{G(\mathcal{B})} \int_{\mathcal{B}} \Gamma(x) \circ G(.)\right).$$

As in the case of the global or local indices Γ and Γ_j of (5.20), (5.19) this index $\Xi(\mathcal{B}, \Gamma)$ does not have an absolute meaning either – it is "only" a tool to compare different fuzzy models for the same fuzzy data or the same process.

5.4.2 Evaluating fuzzy models using probabilistic ideas

A close and transparent analogy can be found in general modeling principles with the aid of statistical means. There it is a necessary step to validate a model and to express the precision attached to it. Most simply, usually, one tests the model with respect to its relevance, and for example an F-test is frequently used. Moreover, the model is equipped with confidence curves associated with the equations of the model. This way of thinking has a long tradition and all its stages are commonly accepted and used in the field of statistical modelling, cf. for example BOX/JENKINS (1970).

The same does not hold true for the fuzzy set approach. Mainly only the first of these steps is merely solved; unfortunately we cannot give a quantitative characterization of the quality (relevance) of a model. A few approaches try to tackle this problem, but in a qualitative way. The proposal formulated here represents an attempt to express and measure the relevance of a model and in its consequences leads to the formation of fuzzy sets of a complex character, namely interval-valued fuzzy sets or fuzzy sets of type 2, i.e. fuzzy sets whose membership degrees are intervals or fuzzy subsets of $[0,1]$; cf. MIZUMOTO/TANAKA (1976), SAMBUC (1975).

In order to provide a clear description of the approach, it will be given in an algorithmic form. This gives a concise presentation and allows potential users to have this idea ready for use. Additionally we include some comments to explain the consecutive steps.

Having at hand a data set represented by pairs (A_i, B_i), $i = 1, \ldots, n$ of fuzzy sets a fuzzy relation is to be constructed to (approximately) solve the corresponding fuzzy relational equations. To focus attention we restrict ourselves to one specified form of the equations, e.g. to (3.26), (3.27). Of course, the procedure described in the following applies to any form of equations as indicated in sections 3.5 and 3.6. The fuzzy relation R is assumed to be obtained via any suitable method.

Let us consider the fuzzy sets $\overline{B}_i$ resulting from the fuzzy sets A_i composed by R, i.e.: $\overline{B}_i = A_i \circ_t R$. If the fuzzy model – here the fuzzy relation R – is perfect, which almost never occurs in practice, then for every $i = 1, \ldots, n$ we have $\overline{B}_i = B_i$. Since this case is not realistic, we may present a global evaluation of the model by computing the (global) equality degree $\overline{B}_i \equiv_t B_i$ for each pair $(\overline{B}_i, B_i)$. But here we are more interested to take the local point of view, i.e. to look "how equal" the fuzzy sets $\overline{B}_i$ and B_i are separate at each point of the universe of discourse $\mathcal{Y}$. Thus we shall refer to the local degrees of equality (5.3) and (5.4). Fixing some $b \in \mathcal{Y}$ we get for $i = 1, \ldots, n$ a sequence of reals indicating how closely the membership degrees $\mu_{B_i}(b)$ and $\mu_{\overline{B}_i}(b)$ are equal to each other. Now, instead of looking for a single number expressing en block a similarity of realized and intended outputs such as (5.5), (5.6), we build a kind of empirical distribution function of some equality degree using as a sample the values

$$\|B_i = \overline{B}_i\|(b) \quad \text{for all } i = 1, \ldots, n. \tag{5.25}$$

Let us denote this function by $F(w; b)$ with $w \in [0,1]$. By definition this distribution function is given as

$$F(w;b) =_{def} \frac{1}{n} \cdot \text{card}\left(\{i \in \{1,\ldots,n\} \mid \|B_i = \overline{B}_i\|(b) \leq w\}\right). \qquad (5.26)$$

In virtue of (5.26), $F(w;b)$ is monotonous in the first argument with $F(1;b) = 1$. We interpret $F(w;b)$ as expressing the probability that the local equality index at b does not exceed the specified value w in any case of the application of the model.[4] Now, let us further introduce the value

$$p = 1 - F(w;b) \qquad (5.27)$$

which will be considered as the probability that our local equality degree at $b \in \mathcal{Y}$ attains values greater than w. Rewriting (5.27) as

$$\text{Prob}\{(\text{equality degree at } b) > w\} = p \qquad (5.28)$$

we recognize a direct dependence between this value of probability p and the length of a kind of confidence interval induced by the value w. For consistency of notation denote this relationship (5.28) by $Q(w;b) = p$. Summarizing we have that p is a nonincreasing function of w.

Obviously, for any fuzzy set B and $w = 0$ we get a confidence interval of length 1, while for $w = 1$ the length of the confidence interval is reduced to zero.

From (5.27) we observe that with increasing values of w we have decreasing values of p. Hence this implies that if p increases, the length of the confidence interval increases as well. Of course, one dislikes accepting confidence intervals which are too broad since this conveys no useful information. On the other hand, confidence intervals which are too narrow cannot be accepted due to the significantly low values of probability attached to them.

From these remarks one can deduce a rationale how to choose a suitable value of w. We want to achieve two contradictory goals. One is to get as high a value of probability as possible. The second is to have confidence intervals which are narrow enough. Denote by $l_b(w)$ the length of this confidence interval. Hence, to fulfill both requirements we consider as a performance index the product

$$Q(w;b) \cdot (1 - l_b(w)). \qquad (5.29)$$

[4]Of course, this idea presupposes that all the control rules ($A \models\!\!\Rightarrow B$) are of the same importance or applied with the same basic probability. That, surely, is the omly reasonable presupposition for a first approach. Nevertheless, a more sophisticated approach should allow one to dispense with this presupposition.

Since the first factor is a nonincreasing function of w and the second one is a nondecreasing one, therefore there is at least one point w_0 for which (5.29) achieves a maximum value:

$$Q(w_0; b) \cdot (1 - l_b(w_0)) = \sup_{w \in [0,1]} Q(w; b) \cdot (1 - l_b(w)). \tag{5.30}$$

This value w_0 might be utilized for the further construction of an interval-valued fuzzy set.

This gradual construction of a value $w_0 = w_0(b)$ is repeated at every point of the universe $\mathcal{Y}$; thus we get a function w_0 with values $w_0(b)$.

In this way the fuzzy model is characterized not only by the fuzzy relation R, but also by a set of functions $Q(w; b)$ – or at least a set of numbers $w_0(b)$, $b \in \mathcal{Y}$, is attached.

For fuzzy relational equations we proceed accordingly. For a given fuzzy set A compute $B = A \circ_t R$ – or having B at hand solve the inverse problem. In both cases a suitable value of w is taken based on a prespecified probability p or of extremal nature as w_0. In the direct mode this strategy yields an interval-valued fuzzy set $[B_-, B_+]$ that gives an impression how precise the fuzzy set is which is "produced" by the fuzzy model behind the system of equations.

For the two ways just described we can give the following interpretations:

- in the first situation we say that the genuine fuzzy set, which appears in a system when A is given as input, has at each point $b \in \mathcal{Y}$ a membership degree with respect to the output lying between $\mu_{B_-}(b)$ and $\mu_{B_+}(b)$;

- for the second construction we have a fuzzy set $[B_-, B_+]$ with highest specificity, thus the bounds $\mu_{B_-}(b)$ and $\mu_{B_+}(b)$ determine a region in $[0, 1]$ in which one is sure to get a value of the membership at a high value of probability and, simultaneously, this interval is chosen to be narrow enough for the actual purposes.

5.4.3 Consequences for fuzzy modelling strategies

Without going deeply into various applications where fuzzy relational equations are discussed we will rather focus on methodological aspects which arise while studying the introduced approach.

For this, note again that having an experimental data set in the form of pairs (A_i, B_i), two models can be constructed and their use may depend on some specifity of the problem discussed. For short let us refer to them as direct and backward use of the intended model. And let us assume that, as before, the data set of pairs (A_i, B_i) determines a system of fuzzy relation equations to be solved to determine the intended model R.

As a schematic illustration of the following ideas, viz. to take two fuzzy models for direct and backward mode of utilization, cf. Figure 5.2.

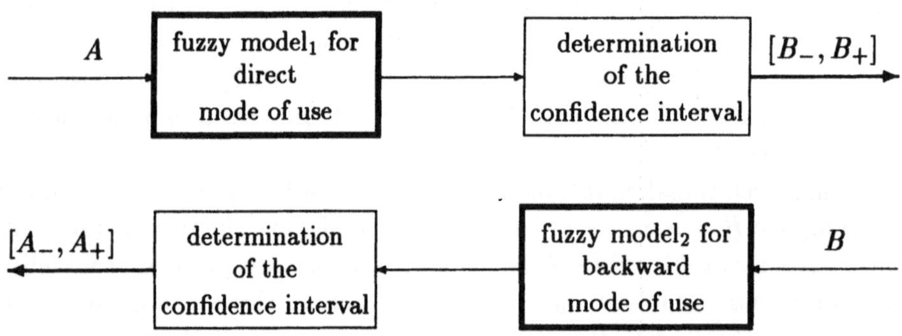

Figure 5.2: Direct and backward use of fuzzy models

In the direct mode of using the model, possessing the fuzzy relation R to tie A and B, viz. B is tied with A via R, we can find the response of the model (system) for any given A straightforwardly.

But in the case that the fuzzy relation has not been determined or that the system of fuzzy relation equations is not solvable and has only approximate solutions the discussion becomes more involved. Thus let us suppose that we only have some confidence intervals indicating the model quality. Following the previously discussed schema of computation of the output we get an interval-valued fuzzy set $[B_-, B_+]$ which expresses bounds in which the response of the system is contained – e.g. with a prescribed level of probability. Afterwards, if one is interested in describing successive states of the system which has a dynamics governed by the constituting fuzzy data, further iterations might be performed. Then the first output is the input for the second step of the iteration. In other words, A in the next step is replaced by $[B_-, B_+]$. For B_- and B_+, separately, the corresponding interval-valued set is built by taking the minimum of the two lower bounds of them and the maximum of their two upper bounds, respectively, as resulting bounds. Obviously, the width of the intervals of this interval-valued fuzzy set is broader than the two original ones on which it is based.

The resulting effect is in agreement with our intuition: the result can never be more precise than the arguments taken into account; cf. as a related paper concerning this topic e.g. CZOGALA/GOTTWALD/PEDRYCZ (1982).

The so-called backward mode of the utilization of the fuzzy model corresponds to all the questions related to A if B is known. Of course, the preferable situation is that R has already been computed. For instance, to answer the question which input A – if any – leads to the output B, we have to solve what is called the inverse problem. Originally, cf. SANCHEZ (1976), assuming a nonempty set of solutions, one immediately started to consider a fuzzy set A fulfilling the model equation, cf. Table 3.2.

But bearing in mind that in most cases the solvability of the basic system of model equations which correspond to the constituting fuzzy data is lacking and one thus has to look at approximate solutions, and that additionally one is confronted with the aspect of model precision in general, it may be useful to reformulate the problem thus:

> *Find A such that the fuzzy set resulting from A is covered by $[B_-, B_+]$, i.e. its membership values are within the bounds of this characteristic interval-valued set $[B_-, B_+]$.*

However, even the solution of this problem might be too tedious a task.

Therefore we can think about a different model just establishing a relation between B and A which is a little different. The model just derived may answer the question which interval-valued fuzzy set $[A_-, A_+]$ is obtained for a specified output fuzzy set B. Surely, the "new" fuzzy relation can be different from the previous one, and the precision of the new fuzzy model might also be different.

As was mentioned previously, here we do not intend to discuss detailed applications.

What remains to be discussed is how the analysis coming from this forward and backward use of models can be enriched by imposing imprecision attached to the interval-valued fuzzy sets. It seems that the forward and backward strategy can transparently be utilized for reasoning schemes in knowledge-based systems with mechanisms managing uncertainty in-built in them. In such a case the models correspond to modelling directions of reasoning.

5.5 Controllability and predictability in fuzzy models

For any process to be modelled by scientific means, it is of interest to know about its controllability, i.e. to know to what extent it is possible to influence and govern – by external choice of some initial or boundary conditions – or even to (completely) control its output. Of course, processes with a very low or even zero level of controllability are without interest for (e.g. computerized) control and for (standard) technical applications in general.

In the same way it is of interest for such processes to have information about their long-range behaviour in time – with respect to either a continuous or a discrete time structure – in order to be able to predict the process behaviour for some longer time interval.

Ideally both these problems should be solved directly for the process under consideration. But for quite different reasons this is not always possible. A kind of way out then is to look at some suitably good model of the process and to discuss these problems of process behaviour not directly for the process but for the model instead. Of course, the results of such an approach have to be interpreted very carefully. But if there is independent information about the model quality, the detour through the model may be an interesting alternative for the direct study of the process.

The process model in particular may be a fuzzy model based on some fuzzy (e.g. linguistic) data concerning the (qualitative) process behaviour. How then can one approach these problems of controllability and predictability? Of course, differing from the usual situation one has in mind when discussing fuzzy controllers, here we suppose that the fuzzy model is a model for the process under consideration in the sense that it models the process in the feedforward manner. The standard situation with fuzzy control, on the contrary, is the "inverse" one: the fuzzy controller is usually supposed to act in the feedback manner with respect to the process and thus already supposes the controllability of the process under consideration and even information on how to control that process.

As usual in our approach we suppose that the fuzzy model is constituted by a fuzzy relation R and that the model's inputs and outputs are connected via the "compositional rule of inference", i.e. via fuzzy relation equations of type (3.2). Therefore our fuzzy model connects with any input data $A \in \mathbb{F}(\mathcal{X})$ the output $B \in \mathbb{F}(\mathcal{Y})$ given by

$$B = R''A = A \circ_t R. \tag{5.31}$$

For the controllability purpose we now make a further restriction: instead of discussing the controllability of the process we discuss the controllability of the model[5]. Therefore the fuzzy output B of the model has to be interpreted as formulating a *goal* to be reached by presenting the model with a suitable input A. In the language of fuzzy equations: now equation (5.31) has to be solved with respect to A for given data R and B (and of course a given t-norm $\boldsymbol{t}$).

From theorem 3.4 (i) we know that in the case of the solvability of an equation of type

$$B = R''X = X \circ_t R \tag{5.32}$$

the greatest solution (with respect to the partial ordering $\subseteq_t$ in $\mathbb{F}(\mathcal{X})$) is given by

$$X = R \downarrow B = \{x \,\|\, \forall y((x,y) \,\varepsilon\, R \rightarrow_t y \,\varepsilon\, B)\}.$$

Furthermore the discussion in section 3.5 adds the information that the fuzzy set $R \downarrow B$ is even the best possible approximate solution of the fuzzy equation (5.32) if there does not exist a solution at all.

Thus independent of the solvability of equation (5.32), for any given goal B the fuzzy set $A = R \downarrow B$ can be taken as an optimal input value for the fuzzy model to reach the goal B at least approximately. Of course, the model output really reached with the input $A = R \downarrow B$ will be the fuzzy set

$$\overline{B} = R''(R \downarrow B) = (R \downarrow B) \circ_t R. \tag{5.33}$$

Again, therefore, the problem arises of comparing the fuzzy sets B and $\overline{B}$.

And again there is the global approach towards that problem which was discussed in the context of the fuzzified equality relation $\equiv_t$ in chapter 3, as well as the localized approach of section 5.4.1 with its idea of a pointwise comparison of B and $\overline{B}$ over the output space $\mathcal{Y}$ of the fuzzy model R.

In the case of the global approach one looks for the truth degree of the formula $B \equiv_t \overline{B}$, i.e. for the truth degree

$$\delta_B =_{def} [\![B \equiv_t R''(R \downarrow B)]\!]. \tag{5.34}$$

[5] Of course, with this decision we accept a – perhaps essential – weakening of the problem of the controllability of the process. The feeling is that this restriction is not too severe if the model has sufficiently high quality. Nevertheless, in any case one needs extra discussion linking the controllability behaviour of the model with the corresponding behaviour of the process.

Using proposition 3.1 (i) one has, assuming LSC(t),

$$\models B \equiv_t R''(R \downarrow B) \leftrightarrow B \subseteqq_t R''(R \downarrow B) \tag{5.35}$$

and thus also

$$\models B \equiv_t R''(R \downarrow B) \rightarrow B \subseteqq_t rg(R) \tag{5.36}$$

by the simple fact that

$$\models R''(R \downarrow B) \subseteqq_t rg(R).$$

Thus there is

$$\delta_B \leq [\![B \subseteqq_t rg(R)]\!] = \inf_{y \in \mathcal{Y}} \sup_{x \in \mathcal{X}} (\mu_B(y) \, \varphi_t \, \mu_R(x, y))$$

which gives quite a simple upper bound of the global degree δ_B to which the goal B can be reached by the fuzzy model R.

The localized approach, for simplicity, will again suppose that the output space $\mathcal{Y} = \{b_1, \ldots, b_m\}$ is a finite set. Then, like in formula (5.20), the actual evaluation vector $\Gamma = (\Gamma_1, \ldots, \Gamma_m)$ will have the components

$$\Gamma_j =_{def} \| B = R''(R \downarrow B) \| (b_j)$$

which means, by the corresponding definitions together with (5.35) and (5.36), that one has

$$\Gamma_j = [\![b_j \varepsilon B \rightarrow_t b_j \varepsilon R''(R \downarrow B)]\!] \leq [\![b_j \varepsilon B \rightarrow_t b_j \varepsilon rg(R)]\!]$$

and which yields for each $j = 1, \ldots, m$ the upper bound

$$\Gamma_j \leq \sup_{x \in \mathcal{X}} (\mu_B(b_j) \, \varphi_t \, \mu_R(x, b_j)).$$

As in section 5.4.1 the localized approach now opens the way for another kind of globalization of the local indices Γ_j which evaluate the controllability of R with respect to the goal B: guided by a fuzzy measure $G : \mathbb{P}(\mathcal{Y}) \rightarrow [0, 1]$ over the output space $\mathcal{Y}$ with the fuzzy integral (5.24) one gets another global controllability index

$$\Lambda_B =_{def} \int_{\mathcal{Y}} \Gamma(x) \circ G(.) = \sup_{\alpha \in [0,1]} \min\{\alpha, G(\Gamma^{\geq \alpha})\} \tag{5.37}$$

for the fuzzy model R with respect to the goal B.

The advantage of this second type of global index (5.37) over the first one (5.34) is again the fact that the fuzzy measure G may distinguish regions in the output space $\mathcal{Y}$ which are more sensitive for the evaluation of the prediction quality of R with respect to the goal B than other ones.

In both cases, however, the evaluation of the prediction quality happens only with respect to a fixed goal B. If one has in mind only some (very) few goals one is interested in reaching by the fuzzy model R this situation may be acceptable. In general, however, one should take into account a second globalization procedure here: a globalization of the prediction quality with respect to "all" the possible goals. In a very abstract sense the class $\mathbb{F}(\mathcal{Y})$ of all fuzzy subsets of the output space $\mathcal{Y}$ is the class of all possible goals. Again now not all these possible goals in $\mathbb{F}(\mathcal{Y})$ will have equal importance. Thus here too a (second) fuzzy measure $\boldsymbol{G} : \mathbb{P}(\mathbb{P}(\mathcal{Y})) \rightarrow [0,1]$ can be taken into consideration to evaluate the different possible goals with respect to their importance for the real control actions one expects. Using this fuzzy measure $\boldsymbol{G}$ the globalization of the evaluation of the controllability of the fuzzy model R with respect to all the possible goals can be given in one of the forms

$$\int_{\mathcal{Z}} \Lambda_B \circ \boldsymbol{G}(.) \qquad \text{or} \qquad \int_{\mathcal{Z}} \delta_B \circ \boldsymbol{G}(.),$$

$\mathcal{Z} = \mathbb{P}(\mathcal{Y})$ the class of all possible goals, depending on which global evaluation of the controllability quality of R with respect to a fixed goal B the approach is based.

Once again, as already done in section 5.4.1, let us however mention that all those evaluation indices (localized or globalized) for the graded controllability property of a fuzzy model R have essentially only a *relative* meaning: they are helpful for comparing either the relative (global) controllability behaviour of different fuzzy models, or the controllability behaviour of one fuzzy model with respect to different goals either in the global or in the localized sense. Their absolute meaning does not usually provide a high level of information.

Besides the controllability problem, we also mentioned initially the predictability problem with respect to a fuzzy model R. To have a closer look at it, we now assume that the fuzzy model under consideration describes the change of an actual situation from one point in time to the next one. That means that we suppose to have a discrete time structure, and also that the input and output spaces of the fuzzy model coincide: $\mathcal{Y} = \mathcal{X}$. Otherwise, in the case that $\mathcal{Y} \neq \mathcal{X}$, the successive outputs through time of the fuzzy model R are determined by the successive inputs the model becomes confronted with – thus all the "prediction" would consist in getting knowledge of the input sequence, but this obviously in general is independent of the fuzzy model under consideration. Thus our restrictions are well motivated.

Given one input fuzzy set $A \in \mathcal{X}$ the sequence of consecutive output fuzzy sets of the process under consideration is denoted by $B_1, B_2, B_3, \ldots$ and the corresponding sequence of consecutive output fuzzy sets of the fuzzy model R by $\overline{B}_1, \overline{B}_2, \overline{B}_3, \ldots$. That means for the fuzzy model that we consider the sequence

$$\begin{aligned} \overline{B}_1 &= R''A, \\ \overline{B}_2 &= R''\overline{B}_1 = R''(R''A) = (R^2)''A, \\ \overline{B}_3 &= R''\overline{B}_2 = R''((R^2)''A) = (R^3)''A, \\ &\vdots \\ \overline{B}_n &= \ldots = (R^{n-1})''A \end{aligned}$$

with the notation R^n for the iterated relational product as introduced in definition 2.13. This sequence $\overline{B}_1, \overline{B}_2, \overline{B}_3, \ldots$ of successive outputs of the fuzzy model R generated by the initial input A is the exact prediction for the *models* behaviour. The predictability for the model is hence not a serious problem.

Thus here we again look at the mutual relation of the sequences of successive outputs of the process on the one hand and of the fuzzy model on the other hand, both starting from the same initial "input situation". What we would like to evaluate now is the property

The fuzzy model R predicts well the systems long range behaviour. (5.38)

Suppose that the accuracy of the fuzzy model R is characterized as before in (5.20) by a vector Γ of indices Γ_j localizing the model quality according to the points of the (finite) input and output space $\mathcal{X} = \{b_1, \ldots, b_m\}$. This vector may be given independently of the present considerations.

Using this vector Γ together with the first output $\overline{B}_1$ of the fuzzy model R and remembering the idea of "confidence intervals" mentioned in sections 5.1 and 5.4.2, we are able to define upper and lower bounds $(B_1)^*$ and $(B_1)_*$ for the first process output B_1. Quite a simple idea is to determine these both bounds pointwise through possible deviations of the membership degrees of $\overline{B}_1$ and B_1.

Having in mind the fact that the localizing indices Γ_j measure some local degree of coincidence of membership degrees and that the negation of such a coincidence degree should be something like a degree for deviation of these membership degrees it is reasonable to use as membership degrees for the fuzzy sets $(B_1)^*$ and $(B_1)_*$ the values

$$\mu_{(B_1)^*}(b_j) = \min\{1, \mu_{\overline{B}_1}(b_j) + (1 - \Gamma_j)\}, \tag{5.39}$$

$$\mu_{(B_1)_*}(b_j) = \max\{0, \mu_{\overline{B}_1}(b_j) - (1 - \Gamma_j)\}. \tag{5.40}$$

In a modified notation more in the traditional style of formal logic, these formulas can be rewritten using the connectives $\&, \uplus$ of the Lukasiewicz many-valued logic as

$$\models\ b_j \,\varepsilon\, (B_1)^* \leftrightarrow b_j \,\varepsilon\, \overline{B}_1 \uplus (\neg \Gamma_j) \leftrightarrow (\Gamma_j \rightarrow_L b_j \,\varepsilon\, \overline{B}_1), \tag{5.41}$$

$$\models\ b_j \,\varepsilon\, (B_1)_* \leftrightarrow \Gamma_j \,\&\, b_j \,\varepsilon\, \overline{B}_1. \tag{5.42}$$

These last mentioned formulations indicate a possible further generalization which will not be discussed here: instead of the Lukasiewicz conjunction & and implication $\rightarrow_L$ some left continuous t-norm $\boldsymbol{t}$ and its corresponding Φ-operator $\varphi_{\boldsymbol{t}}$ could be taken. The crucial point with such a generalization, however, seems to be that one first has to check in which sense other t-norms than $\boldsymbol{t}_L$ adapt the idea of "confidence intervals" which was constitutive for the approach through (5.39), (5.40).

Unfortunately, however, this rewriting does not yield "nice" formulas. Therefore we additionally look for a more set theoretical shape for (5.39) and (5.40). To reach that goal we have to reconsider the evaluation vector $\Gamma = (\Gamma_1, \ldots, \Gamma_m)$ of the local comparison indices. The idea leading to the introduction of such an evaluation vector Γ in (5.20) was to connect with each point $b_j \in \mathcal{Y}$ its local comparison index $\Gamma_j \in [0,1]$, therefore Γ is nothing other than a fuzzy subset of the input and output space $\mathcal{X} = \mathcal{Y}$. For the membership degrees one obviously has

$$\mu_\Gamma(b_j) = \Gamma_j.$$

Rereading (5.41), (5.42) with this interpretation of Γ in mind immediately gives the simple representations

$$(B_1)^* = \complement\Gamma \cup_{\boldsymbol{t}_L} \overline{B}_1 = \Gamma \rhd_{\boldsymbol{t}_L} \overline{B}_1 \quad \text{and} \quad (B_1)_* = \Gamma \cap_{\boldsymbol{t}_L} \overline{B}_1. \tag{5.43}$$

With these fuzzy sets $(B_1)^*$ and $(B_1)_*$ we now have an upper and a lower bound for the process output B_1 to the input A in the sense that we expect to have

$$\models\ (B_1)_* \subseteqq_{\boldsymbol{t}} B_1 \subseteqq_{\boldsymbol{t}} (B_1)^*. \tag{5.44}$$

Of course, both bounds $(B_1)_*, (B_1)^*$ have to be functions simply of the input A (and the fuzzy model R). This, together with the idea that the construction of these bounds for B_1 will become suitably iterated to also discuss bounds for the further process outputs $B_2, B_3, \ldots$ forces us to introduce an even more flexible notation. Hence we define for any $A \in \mathbb{F}(\mathcal{X})$

$$\mathsf{U}_R(A) \;=_{def}\; \mathbf{C}\Gamma \cup_{\mathbf{t}_L} R''A, \tag{5.45}$$

$$\mathsf{L}_R(A) \;=_{def}\; \Gamma \cap_{\mathbf{t}_L} R''A. \tag{5.46}$$

Independent of the idea behind (5.44) one immediately has for each fuzzy set $A \in \mathbb{F}(\mathcal{X})$:

$$\models \mathsf{L}_R(A) \subseteqq_{\mathbf{t}} \mathsf{U}_R(A). \tag{5.47}$$

These upper and lower bounds (5.45) and (5.46) provide the formal basis for the problem of the long range prediction of the process behaviour. In GOTTWALD/PEDRYCZ (1986a) a worst case discussion used them in the sense that the bounds for the next step were taken in such a way that the new lower bound was the minimum of the lower bounds of the model outputs for both previous bounds and the new upper bound was the maximum of the upper bounds of the same model outputs. This strategy, however, was unnecessarily difficult and can be simplified just by using some monotonicity properties of the operators $\mathsf{U}_R, \mathsf{L}_R$ which generalize property (5.47).

Proposition 5.1 *For each fuzzy model $R \in \mathbb{F}(\mathcal{X} \times \mathcal{X})$, all fuzzy sets $A, B \in \mathbb{F}(\mathcal{X})$ and each t-norm $\mathbf{t}$ with property $\mathsf{LSC}(\mathbf{t})$ there hold true*

(*i*) if $\models A \subseteqq_{\mathbf{t}} B$ then $\models \mathsf{U}_R(A) \subseteqq_{\mathbf{t}} \mathsf{U}_R(B)$,

(*ii*) if $\models A \subseteqq_{\mathbf{t}} B$ then $\models \mathsf{L}_R(A) \subseteqq_{\mathbf{t}} \mathsf{L}_R(B)$.

Proof. Straightforward from the corresponding definitions or, even better, by reference to theorem 2.1. QED

Compared with the usual style of results on fuzzy sets we intended to prove in chapter 2, these monotonicity results are weak. Nevertheless, they suffice for the present considerations on the prediction problem. But indeed, they can be generalized. Unfortunately,howevert, in order to do so we need some cumbersome assumptions, resulting from a reference to proposition 2.2 in the proof of these generalizations. It was mainly for this reason that the more restricted proposition 5.1 was presented first.

Proposition 5.2 *For each fuzzy model $R \in \mathbb{F}(\mathcal{X} \times \mathcal{X})$, all fuzzy sets $A, B \in \mathbb{F}(\mathcal{X})$ and each t-norm $\mathbf{t}$ with property $\mathsf{LSC}(\mathbf{t})$ there hold true*

(*i*) $\models A \subseteqq_{\mathbf{t}} B \rightarrow \mathsf{L}_R(A) \subseteqq_{\mathbf{t}} \mathsf{L}_R(B)$,

and if the t-norm $\mathbf{t}$ also distributes over the Lukasiewicz disjunction $\uplus$ or fulfills one of the conditions (2.16), (2.17) with respect to $\mathbf{s}_{\mathbf{t}_1} = \uplus$ there also holds true

(*ii*) $\models A \subseteqq_{\mathbf{t}} B \rightarrow \mathsf{U}_R(A) \subseteqq_{\mathbf{t}} \mathsf{U}_R(B)$.

Proof. (i) By proposition 2.18 (i) one has

$$\models A \subseteqq_{\mathbf{t}} B \rightarrow R''A \subseteqq_{\mathbf{t}} R''B$$

and by proposition 2.2 (ii) then also

$$\models R''A \subseteqq_{\mathbf{t}} R''B \rightarrow \mathbf{C}\Gamma \cup_{\mathbf{t}_L} R''A \subseteqq_{\mathbf{t}} \mathbf{C}\Gamma \cup_{\mathbf{t}_L} R''B.$$

Using the transitivity of $\subseteqq_{\mathbf{t}}$ together this yields (i).

(ii) is proven in the same manner but with reference to proposition 2.2 (iv) instead of proposition 2.2 (ii). And this reference together with the remark following proposition 2.2, causes the additional assumptions for (ii). QED

After this theoretical side remark concerning the monotonicity of the operators $\mathsf{U}_R, \mathsf{L}_R$, let us return to the prediction problem. Already the (weaker) monotonicity results of proposition 5.1 allows one to simplify the previously mentioned discussion of GOTTWALD/PEDRYCZ (1986a) equivalently in considering as upper bounds for the process outputs $B_1, B_2, B_3, \ldots$ the successive iterations

$$\begin{aligned} \mathsf{U}^1(A) &= \mathsf{U}(A), \\ \mathsf{U}^2(A) &= \mathsf{U}(\mathsf{U}(A)), \\ \mathsf{U}^3(A) &= \mathsf{U}(\mathsf{U}^2(A)), \\ &\vdots \end{aligned}$$

of the upper bound operator (5.45) and as lower bounds of the process outputs the corresponding iterations of the lower bound operator (5.46):

$$\begin{aligned} \mathsf{L}^1(A) &= \mathsf{L}(A), \\ \mathsf{L}^2(A) &= \mathsf{L}(\mathsf{L}(A)), \\ \mathsf{L}^3(A) &= \mathsf{L}(\mathsf{L}^2(A)), \\ &\vdots \end{aligned}$$

To get now information on the prediction quality of the fuzzy model R (with respect to the initial input A) for a fixed, say k-step time horizon, we compare the bounds $\mathsf{U}^k(A)$ and $\mathsf{L}^k(A)$ in the localized manner of sections 5.1 and 5.4.1. That means we regard for each point $b_j \in \mathcal{X} = \{b_1, \ldots, b_m\}$ the index

$$\eta_j^k =_{def} \|\mathsf{U}^k(A) = \mathsf{L}^k(A)\|(b_j)$$

which can be seen as defining a fuzzy subset η^k of $\mathcal{X}$ whose membership degrees are always $\mu_{\eta^k}(b_j) = \eta_j^k$.

Again now a globalization is possible of these local evaluations of the prediction quality of R for the time horizon k. As before we suppose that a fuzzy measure G over $\mathcal{X}$ is given such that using the fuzzy integral the globalization happens with respect to this fuzzy measure G and yields the index

$$\Psi(R, k) =_{def} \int_{\mathcal{X}} \eta^k(x) \circ G(.)$$

of k-step predictability.

Taken in an absolute sense, for a fixed fuzzy model R this index $\Psi(R, k)$ simply indicates the decline in prediction quality with a growing length of the time horizon. That is nothing which is new or of deeper interest and only a kind of formal confirmation of the well-known fact that the prediction quality decreases with longer prediction intervals. The real interest in the index $\Psi(R, k)$ therefore is again in its use to compare different fuzzy models of the same process with respect to their prediction behaviour.

Bibliography

Bandemer, H. and S. Gottwald

(1989) *Einführung in Fuzzy-Methoden.* Theorie und Anwendung unscharfer Mengen. Berlin (Akademie-Verlag). [3rd ed. 1992]

Bellman, R. and M. Giertz

(1973) On the analytic formalism of the theory of fuzzy sets. *Information Sci.* **5**, 149 – 156.

Box, G. E. P. and G. M. Jenkins

(1970) *Time Series Analysis.* Forecasting and Control. San Francisco (Holden Day).

Chakraborty, M. K. and M. Das

(1983) Studies in fuzzy relations over fuzzy subsets. *Fuzzy Sets Syst.* **9**, 79 – 89.

(1983a) On fuzzy equivalence. I, II. *Fuzzy Sets Syst.* **11**, 185 – 193 and 299 – 307.

Chang, C. C. and H. J. Keisler

(1973) *Model Theory.* Amsterdam (North-Holland Publ. Comp.).

Chapin, E. W.

(1974/75) Set-valued set theory. I, II. *Notre Dame J. Formal Logic* **15**, 614 – 634; **16**, 255 – 267.

Corge, Ch.

(1975) *Elements d'Informatique.* Informatique et Demarche de l'Esprit. Paris (Larousse).

Czogala, E.; J. Drewniak and W. Pedrycz

(1982) Fuzzy relation equations on a finite set. *Fuzzy Sets Syst.* **7**, 89 – 101.

Czogala, E.; S. Gottwald and W. Pedrycz

(1982) Aspects for the evaluation of decision situations. In: *Fuzzy Information and Decision Processes* (M. M. Gupta, E. Sanchez, eds.), Amsterdam (North-Holland Publ. Comp.), 41 - 49.

Czogala, E. and W. Pedrycz

(1981) Some problems concerning the construction of algorithms of decision-making in fuzzy systems. *Intern. J. Man-Machine Stud.* **15**, 201 - 211.

Di Nola, A.; W. Pedrycz and S. Sessa

(1985) On measures of fuzziness of solutions of fuzzy relation equations with generalized connectives. *J. Math. Anal. Appl.* **121**, 443 - 453.

Di Nola, A. and S. Sessa

(1983) On the set of solutions of composite fuzzy relation equations. *Fuzzy Sets Syst.* **9**, 275 - 285.

Di Nola, A.; S. Sessa; W. Pedrycz and E. Sanchez

(1989) *Fuzzy Relation Equations and Their Applications to Knowledge Engineering.* Theory and Decision Libr., ser. D, Dordrecht (Kluwer Acad. Publ.).

Dombi, J.

(1982) A general class of fuzzy operators, the de Morgan class of fuzzy operators and fuzziness measures induced by fuzzy operators. *Fuzzy Sets Syst.* **8**, 149 - 163.

Dubois, D. and H. Prade

(1980) *Fuzzy Sets and Systems: Theory and Applications.* New York (Academic Press).

Frank, H. J.

(1979) On the simultaneous associativity of $F(x,y)$ and $x + y - F(x,y)$. *Aequat. Math.* **19**, 194 - 226.

Giles, R.

(1976) Lukasiewicz logic and fuzzy set theory. *Intern. J. Man-Machine Stud.* **8**, 313 - 327.

(1979) A formal system for fuzzy reasoning. *Fuzzy Sets Syst.* **2**, 233 - 257.

Gödel, K.

(1932) Zum intuitionistischen Aussagenkalkül. *Anzeiger Akad. Wiss. Wien*, math.-naturwiss. Kl., **69**, 65 – 66.

Gottwald, S.

(1971) Zahlbereichskonstruktionen in einer mehrwertigen Mengenlehre.
Ztschr. math. Logik Grundl. Math. **17**, 145 – 188.

(1971a) Elementare Inhalts- und Maßtheorie in einer mehrwertigen Mengenlehre. *Math. Nachr.* **50**, 27 – 68.

(1974) Mehrwertige Anordnungsrelationen in klassischen Mengen. *Math.*
Nachr. **63**, 205 – 212.

(1976) A cumulative system of fuzzy sets. In: *Set Theory and Hierarchy Theory.* Memorial Tribute A. Mostowski, Bierutowice 1975 (A. Zarach et al., eds.), Lecture Notes Math., vol. 537, Berlin (Springer), 109 – 119.

(1979) Set theory for fuzzy sets of higher level. *Fuzzy Sets Syst.* **2**, 125 – 151.

(1979a) Eine Anwendungsvariante der mehrwertigen Logik. *Wiss. Ztschr. KMU Leipzig*, Ges.- u. Sprachwiss. R., **28**, 303 – 312.

(1980) Fuzzy uniqueness of fuzzy mappings. *Fuzzy Sets Syst.* **3**, 49 – 74.

(1983) Generalization of some results of Elie Sanchez. *BUSEFAL* (Laborat. LSI, Univ. Paul Sabatier, Toulouse), no. **16**, 54 – 60.

(1984) T-Normen und φ-Operatoren als Wahrheitswertfunktionen mehrwertiger Junktoren. In: *Frege Conference 1984*, Proc. Intern. Conf. Schwerin Sept. 10–14, 1984 (G. Wechsung, ed.), Math. Research, vol. 20, Berlin (Akademie-Verlag), 121 – 128.

(1984a) Criteria for non-interactivity of fuzzy logic controller rules. In: *Large Scale Systems: Theory and Applications*, Proc. 3rd IFAC/IFORS Symp. Warsaw 1983 (A. Straszak, ed.), Oxford (Pergamon Press), 229 – 233.

(1984b) On the existence of solutions of systems of fuzzy equations. *Fuzzy Sets Syst.* **12**, 301 – 302.

(1984c) Fuzzy set theory. Some aspects of the early development. In: *Aspects of Vagueness* (H.-J. Skala, S. Termini, E. Trillas, eds.), Theory and Decision Libr., vol. 39, Dordrecht (Reidel), 13 – 29.

(1986) Characterizations of the solvability of fuzzy equations. *Elektron. Informationsverarb. Kybernet.* EIK **22**, 67 – 91.

(1986a) Fuzzy set theory with t-norms and φ-operators. In: *The Mathematics of Fuzzy Systems* (A. Di Nola, A. G. S. Ventre, eds.), Interdisciplinary Systems Res., vol. 88, Köln (TÜV Rheinland), 143 – 195.

(1986b) On some theoretical problems concerning the construction of fuzzy controllers. In: *Fuzzy Sets Applications, Methodological Approaches, and Results* (St. F. Bocklisch et al., eds.), Math. Research, vol. 30, Berlin (Akademie-Verlag), 45 – 55.

(1989) *Mehrwertige Logik*. Eine Einführung in Theorie und Anwendungen. Berlin (Akademie-Verlag).

(1990) Some observations and problems connected with fuzzy relation equations. *Fasciculi Mathematici*, Nr. **19**, Poznan (Polytech. Poznan. Inst. Math.), 87 – 92.

(1991) Fuzzified fuzzy relations. In: *Proc. IFSA '91 Brussels* (R. Lowen, M. Roubens, eds.), vol.: Mathematics, Brussels (Vrije Univ. Brussels), 82 – 86.

(1992) On t-norms which are related to distances of fuzzy sets. *BUSEFAL*, no. **50**, 25 - 30.

Gottwald, S. and W. Pedrycz

(1985) Analysis and synthesis of fuzzy controller. *Problems Control Inform. Theory* **14**, 33 – 45.

(1986) Solvability of fuzzy relational equations and manipulation of fuzzy data. *Fuzzy Sets Syst.* **18**, 1 – 21.

(1986a) On the suitability of fuzzy models: an evaluation through fuzzy integrals. *Intern. J. Man-Machine Stud.* **24**, 141 – 151.

(1988) On the methodology of solving fuzzy relational equations and its impact on fuzzy modelling. In: *Fuzzy Logic in Knowledge-Based Systems, Decision and Control* (M. M. Gupta, T. Yamakawa, eds.), Amsterdam (North-Holland Publ. Comp.), 197 – 210.

Hamacher, H.

(1978) *ber logische Aggregationen nicht-binär explizierter Entscheidungskriterien*. Frankfurt/Main (Rita G. Fischer Verlag).

Hirota, K. and W. Pedrycz

(1983) Analysis and synthesis of fuzzy systems by the use of probabilistic sets. *Fuzzy Sets Syst.* **10**, 1 – 13.

Holmblad, L. P. and J. J. Østergaard

(1982) Control of a cement kiln by fuzzy logic. In: *Fuzzy Information and Decision Processes* (M. M. Gupta, E. Sanchez, eds.), Amsterdam (North-Holland Publ. Comp.), 389 – 399.

Kaufmann, A.

(1975) *Introduction to the Theory of Fuzzy Subsets.* Vol. 1: *Fundamental Theoretical Elements.* New York (Academic Press).

(1977) *Introduction à la théorie des sous-ensembles flous.* A l'usage des
ingénieurs. t. 4: *Compléments et nouvelles applications.* Paris (Masson).

Klaua, D.

(1964) *Allgemeine Mengenlehre.* Berlin (Akademie-Verlag).

(1970) Stetige Gleichmächtigkeiten kontinuierlich-wertiger Mengen. *Monatsber. Deut. Akad. Wiss. Berlin* **12**, 749 – 758.

Klement, E. P.

(1982) Construction of fuzzy σ-algebras using triangular norms. *J. Math. Anal. Appl.* **85**, 543 – 565.

Li H.-X.

(1985/86) Fuzzy perturbation analysis. Part I: Directional perturbation. *Fuzzy Sets Syst.* **17**, 189 – 197;
Part II: Undirectional perturbation, *ibid.* **19**, 165 – 175.

Ling, C. H.

(1965) Representation of associative functions. *Publ. Math. Debrecen* **12**, 182 – 212.

Łukasiewicz, J.

(1970) *Selected Works* (L. Borkowski, ed.). Amsterdam (North-Holland Publ. Comp.).

Łukasiewicz, J. and A. Tarski

(1930) Untersuchungen über den Aussagenkalkül. *Comptes Rendus Soc. Sci. et Lettr. Varsovie,* cl. III, **23**, 30 – 50.

Mamdani, E. H.

(1974) Application of fuzzy algorithms for the control of a simple dynamic plant. *Proc. IEEE* **121**, 1585 – 1588.

(1976) Advances in the linguistic synthesis of fuzzy controllers. *Intern. J. Man-Machine Stud.* **8**, 669 – 678.

Mamdani, E. H. and S. Assilian

(1975) An experiment in linguistic synthesis with a fuzzy logic controller. *Intern. J. Man-Machine Stud.* **7**, 1 – 13. [cf. also Mamdani/Gaines (1981)]

Mamdani, E. H. and B. R. Gaines, eds.

(1981) *Fuzzy Reasoning and Its Applications.* New York (Academic Press).

Mizumoto, M.

(1982) Fuzzy inference using max-$\wedge$-composition in the compositional rule of inference. In: *Approximate Reasoning in Decision Analysis* (M. M. Gupta, E. Sanchez, eds.), Amsterdam (North-Holland Publ. Comp.), 67 – 76.

(1989) Pictorial representations of fuzzy connectives, part I: Cases of t-norms, t-conorms and averaging operators. *Fuzzy Sets Syst.* **31**, 217 – 242.

Mizumoto, M. and K. Tanaka

(1976) Some properties of fuzzy sets of type 2. *Information and Control* **31**, 312 – 340.

(1979) Some properties of fuzzy numbers. In: *Advances in Fuzzy Set Theory and Applications* (M. M. Gupta, R. K. Ragade, R. R. Yager, eds.), Amsterdam (North-Holland Publ. Comp.), 153 – 164.

Mizumoto, M. and H.-J. Zimmermann

(1982) Comparison of fuzzy reasoning methods. *Fuzzy Sets Syst.* **8**, 253 – 283.

Ohsato, A. and T. Sekiguchi

(1983) Convexly combined form of fuzzy relational equations and its application to knowledge representation. In: *Proc. Intern. Conf. Systems, Man and Cybernet., Bombay 1983/84;* vol. 1, Bombay – New Delhi (IEEE India Council), 294 – 299.

(1985) Maximin solution of the convexly combined form of composite fuzzy relation equations [in Japanese]. *Transact. Soc. Instrument and Control Engineers* (Japan) **21**, 423 – 428.

Ovchinnikov, S. V.

(1981) Structure of fuzzy binary relations. *Fuzzy Sets Syst.* **6**, 169 – 195.

Pedrycz, W.

(1982) *Fuzzy control and fuzzy systems.* Dept. Math., Delft Univ. of Technology, Report 82 14.

(1983) Fuzzy relational equations with generalized connectives and their applications. *Fuzzy Sets Syst.* **10**, 185 – 201.

(1983a) Numerical and applicational aspects of fuzzy relational equations. *Fuzzy Sets Syst.* **11**, 1 – 18.

(1985) Applications of fuzzy relational equations for methods of reasoning in presence of fuzzy data. *Fuzzy Sets Syst.* **16**, 163 - 175.

(1988) Approximate solutions of fuzzy relational equations. *Fuzzy Sets Syst.* **28**, 183 - 202.

(1989) *Fuzzy Control and Fuzzy Systems.* Taunton - New York (Research Stud. Press - Wiley).

(1990) Relevancy of fuzzy models. *Information Sci.* **52**, 285 - 302.

(1990a) Direct and inverse problem in comparison of fuzzy data. *Fuzzy Sets Syst.* **34**, 223 - 235.

Pedrycz, W.; E. Czogala and K. Hirota

(1984) Some remarks on the identification problem in fuzzy systems. *Fuzzy Sets Syst.* **12**, 185 - 189.

Rescher, N.

(1969) *Many-Valued Logic.* New York (McGraw-Hill).

Rodabaugh, E. S.; E. P. Klement and U. Höhle, eds.

(1992) *Applications of Category Theory to Fuzzy Subsets.* Dordrecht (Kluwer Acad. Publ.).

Sanchez, E.

(1974) *Equations de relations floues.* Thèse de Doctorat, Faculté de Médecine de Marseille.

(1976) Resolution of composite fuzzy relation equations. *Information and Control* **30**, 38 - 48.

(1977) Solutions in composite fuzzy relation equations: application to medical diagnosis in Brouwerian logic. In: *Fuzzy Automata and Decision Processes* (M. M. Gupta, G. N. Saridis. B. R. Gaines, eds.), Amsterdam (North-Holland Publ. Comp.), 221 - 234.

(1978) Resolution of eigen fuzzy sets equations. *Fuzzy Sets Syst.* **1**, 69 - 74

(1984) Solution of fuzzy equations with extended operations. *Fuzzy Sets Syst.* **12**, 237 - 248.

Schweizer, B. and A. Sklar

(1961) Associative functions and statistical triangle inequalities. *Publ. Math. Debrecen* **8**, 169 - 186.

(1983) *Probabilistic Metric Spaces.* Amsterdam (North-Holland Publ. Comp.).

Sugeno, M.

(1974) *Theory of Fuzzy Integral and Its Applications.* Ph. D. Thesis, Tokyo Inst. of Technology, Tokyo.

(1977) Fuzzy measures and fuzzy integrals: a survey. In: *Fuzzy Automata and Decision Processes* (M. M. Gupta, G. N. Saridis, B. N. Gaines, eds.), Amsterdam (North-Holland Publ. Comp.), 89 - 102.

Thole, U.; H.-J. Zimmermann and P. Zysno

(1979) On the suitability of minimum and product operators for the intersection of fuzzy sets. *Fuzzy Sets Syst.* **2**, 167 - 180.

Wagenknecht, M. and K. Hartmann

(1986) On the solution of direct and inverse problems for fuzzy equation systems with tolerances. In: *Fuzzy Sets Applications, Methodological Approaches, and Results* (St. Bocklisch et al., eds.); Math. Research, vol. 30, Berlin (Akademie-Verlag), 37 - 44.

(1986a) Fuzzy modelling with tolerances. *Fuzzy Sets Syst.* **20**, 325 - 332.

Weber, S.

(1983) A general concept of fuzzy connectives, negations and implications based on t-norms and t-conorms. *Fuzzy Sets Syst.* **11**, 115 - 134.

Weidner, A. J.

(1981) Fuzzy sets and Boolean-valued universes. *Fuzzy Sets Syst.* **6**, 61 - 72.

Yager, R. R.

(1979) A measurement-informational discussion of fuzzy union and fuzzy intersection. *Intern. J. Man-Machine Stud.* **11**, 189 - 200.

(1980) On a general class of fuzzy connectives. *Fuzzy Sets Syst.* **4**, 235 - 242.

Zadeh, L. A.

(1965) Fuzzy sets. *Information and Control* **8**, 338 - 353.

(1971) Similarity relations and fuzzy orderings. *Information Sci.* **3**, 159 - 176.

(1971a) Toward a theory of fuzzy systems. In: *Aspects of Network and System Theory* (R. E. Kalman, N. de Claris, eds.), New York (Holt, Rinehart and Winston), 469 - 490.

(1973) Outline of a new approach to the analysis of complex systems and decision processes. *IEEE Trans. Systems, Man and Cybernet.* **SMC-3**, 28 – 44 [cf. also (1987)].

(1975) The concept of a linguistic variable and its application to approximate reasoning. I. - III. *Information Sci.* **8**, 199 – 250 and 301 – 357; **9**, 43 – 80. [cf. also (1987)]

(1978) Fuzzy sets as a basis for a theory of possibility. *Fuzzy Sets Syst.* **1**, 3 – 28. [cf. also (1987)]

(1983) The role of fuzzy logic in the management of uncertainty in expert systems. *Fuzzy Sets Syst.* **11**, 199 – 227. [cf. also (1987)]

(1984) A theory of commonsense knowledge. In: *Aspects of Vagueness* (H. J. Skala, S. Termini, E. Trillas, eds.), Dordrecht (Reidel), 257 – 295. [cf. also (1987)]

(1987) *Fuzzy Sets and Applications.* Selected Papers. (R. R. Yager et al., eds.), New York (Wiley).

Zhang, J.-W.

(1980) A unified treatment of fuzzy set theory and Boolean-valued set theory – fuzzy set structures and normal fuzzy set structures. *J. Math. Anal. Appl.* **76**, 297 – 301.

Zimmermann, H.-J.

(1985) *Fuzzy Set Theory – and Its Applications.* Dordrecht (Kluwer - Nijhoff). [2nd ed. 1991]

Zimmermann, H.-J. and P. Zysno

(1980) Latent connectives in human decision making. *Fuzzy Sets Syst.* **4**, 37 – 51.

Index

S

T

U

V

Z